A Solar Manifesto

SECOND EDITION

A Solar Manifesto

Hermann Scheer

Published by James & James (Science Publishers) Ltd
35–37 William Road, London, NW1 3ER, UK

A catalogue record for this book is available from the British Library.

ISBN 1 902916 24 7

Printed in the UK by The Cromwell Press

Contents

Introduction

Politics at the Turning Point

A 'Prohibition of military or any other hostile use of environmental modification techniques' has been part of international law since 1977. According to Article 1, each signatory state ('State Party to this Convention') pledges

> not to engage in military or any other hostile use of environmental modification techniques having wide-spread, long-lasting or severe effects as the means of destruction, damage or injury to any other State Party.

But at the same time the world's own civil war against the environment, not prohibited by any treaty, is escalating. It is waged, entirely unilaterally, against forces that are not enemies but rather the elements of life for man. Periods of drought, devastating storms, expanding desert areas, flood disasters, the hole in the ozone layer, are signs of catastrophes that had not been expected to unfold until the medium or long term.

Even the United Nations Conference on the Environment and Development in June 1992 in Rio de Janeiro – the 'Earth Summit' – did not bring about any political change. The conference was based on the available and accepted scientific findings about the destruction of nature, a topic that for at least two decades has demanded fundamental political and economic reorientation. Even more alarming is the recognition that political decision-makers are obviously not capable of such a reorientation. For instance, they managed to agree on a convention to stabilize CO_2 emissions by the year 2000, but even this is not designed to protect the atmosphere; it is merely a 'compromise' to prevent a further increase in the danger level. In fact, it allows the current destruction of the earth's atmosphere to continue. The parties were not even capable of guaranteeing that this small outcome will be adhered to. The terrifying gulf between our knowledge of these present threats and taking the political action to fend them off has not been reduced since then. Obviously, the key players and decision-makers of the world have become so embroiled in the labyrinth of traditional patterns of compulsive action that they can no longer conceive of a way out.

This was confirmed in the five years from 1992 onwards. Man's energy consumption accelerated rapidly. When the governments met in December 1997 at the World Climate Conference at Kyoto in Japan, they were under public pressure to reach a binding agreement for all participants to reduce greenhouse gases. At the last minute this was achieved. EU countries undertook to reduce emissions by 8%, the USA by 7% and Japan by 6% by the year 2010, starting from the 1990 levels. It remains to be seen whether a sufficient number of states will subscribe to this obligation, and whether their parliaments will ratify the agreement. Then again, it is questionable whether and how it will be adhered to. Japan has now declared that it is going to develop further its nuclear power plants. Many people besides the nuclear lobby in Japan are hoping for a renaissance of nuclear energy instead of targeting renewable energy as the non-fossil fuel alternative to atomic fission or fusion.

This is not surprising. The same individuals who, because of their political and economic strategies, are responsible for the destruction of nature are hardly in a position to act as suitable instigators of a new alternative. It is impossible for those who create threats in the first place to act against their own interests and come up with radical political decisions. Above all, the global environmental crisis is a global crisis of politics and policies.

It is not necessary, nor is it the intention of this book, to add yet another volume to the reports already available about the warming of the earth's atmosphere, the radioactive contamination of entire regions, the poisoning and erosion of soils, the contamination of bodies of water, air pollution, the burning of vegetation, the extermination of biological species, or the depletion of the ozone layer. All the new evidence being added to the body of scientific knowledge indicates more and more that the actual degree of damage is higher than assumed so far, and that its consequences will afflict us faster than expected.

Today, more than ever before, there is the utmost urgency for answers to the question of why there are no political strategies, long overdue, to achieve peace with nature.

Conference follows conference, but more often than not the result is nothing more than a call for yet another follow-up conference. Report follows report – and the result is usually a contract for yet another report. Political decision-makers reassure and excuse themselves by pointing to the shared responsibility of others without whom, they say, they can do nothing. Instead of changing their own strategies, they preach a change in general consciousness. Appeals are made to rethink the issues – especially by those with a mandate to act. Decision-makers divert attention from their own responsibilities by shifting them to the general public. The threat is acknowledged in theory but ignored in practice. Another decade of anxiety-laden speeches, reports and conferences as substitutes for political action would be life-threatening for mankind.

But why do we stick to the policy of only marginal practical value that proclaims 'Peace with Nature', when those perennial 'shocked, shocked' speeches should have caused a fundamental change of priorities long ago? Why is it that presumably responsible governments, parliaments and international organizations have managed only to devise countermeasures that, even if successful, are nothing more than the metaphoric drop in the bucket, a single drip on the rapidly heating-up globe? Why are they risking collective doom despite the fact that they have all the relevant information and data at their fingertips? Why do those who are in positions to do something effective not fight back? Why is there lethargy instead of a popular uprising? Have more and more people resigned themselves to the brewing disaster and written off the future? What explains the strange harmony of politically misdirected decisions in the face of exploding contradictions? Never before have there been better, or more urgent, reasons for fundamental innovation, but there is a basic lack of new activist supporters. Why has it been impossible for any political organization to develop an appropriate and practicable strategy? Who will be the true supporters of such a process? Mankind's fate depends on finding the right answers to these questions.

The dangers of destroying the environment, and with it civilization, are caused not by a lack of knowledge and intelligence in political and economic leaders, but by their stubborn refusal to find a way out of the fossilized structures of thought and action that direct their programmes, and on which their very existence depends. They are continuously seeking solutions within the existing frames of reference, which they believe could be ecologically *supplemented* or *modified* at best, but which, in their view, should by no means be *replaced*. With this approach, they practise surgery on symptoms, and become increasingly less successful the longer they restrict themselves to that approach. No longer do they make any attempt to find a common thread that connects these problems, because they have given up questioning the connection itself. And, since the collapse of the Eastern Bloc, the West's leaders, in their self-deceiving euphoria of victory, see less reason for that than ever before.

The need for a change in political paradigms by the end of this century of environmental destruction has been urged repeatedly – for example, in Fritjof Capra's book published in the 1970s, *The Turning Point*.[1] There are comprehensive analytical arguments for such pressing changes, but a lack of corresponding practical strategic designs. The philosopher Hans Jonas, warning voice of the 'Responsibility Principle', in an interview on the occasion of the Rio Conference, reached the resigned conclusion that he still had not given up hope that mankind would yet learn how to change its way of thinking.[2] The German social philosopher Carl Amery, in his book *Nature as Politics*, concludes that the logic of mankind's survival dictates the 'fastest possible destruction of the industrial system, at any price'.[3] The German New Left author Robert Kurz, in his book *The Collapse of Modernization*, concludes that it makes no sense 'in view of the collective suicidal actions on a global scale even to discuss individual "reforms" as long as this discussion does not lead to a perspective of the genuine abolition of modern goods and their global system'.[4]

The great political drama of the present age is that unvarnished analyses such as these and their resulting conclusions have not yet produced strategies that genuinely correspond to the depth and width of the real problems – those that must be tackled and realized, that must be visualized and experienced, and whose immediate advantages can be demonstrated and mobilized for mankind's long-term benefit.

More and more people are reaching the conclusion that collective self-destruction can no longer be avoided, and that we are, irreversibly, in a process of decline. In contrast, the conceptual and intellectual *leitmotiv* of this book is that this descent can be reversed – although at present there are many more reasons against this thesis than there are supporting it. The essential precondition is a *change of paradigms of political and economic strategies* – in other words, a basic realignment of standards of action; a political revolution of a type and quality without precedent. But equally unprecedented is the danger to be overcome.

From Agenda 21 to Agenda 1: Solar Energy

In spite of its lack of binding obligation, Agenda 21 of the Rio Conference has become a document of new international political benchmarks for those who are taking the problem seriously. The document makes it possible to demonstrate the gulf between words and deeds, and between acknowledgement of the problems and suitable lasting practical actions. However, Agenda 21 also shows us how even those who are striving for a solution to the crisis end up making strategic omissions.

Agenda 21 listed the basic themes – a catalogue of planetary ills for the 21st century – that were the focal points of the conference: climate change, destruction of the ozone layer, cross-border air pollution, soil erosion, desertification, loss of biological diversity, biotechnological risks, pollution of water resources, environmental damage due to shipping, pollution of oceans and coastal areas, destruction of marine life, and chemical, hazardous and solid waste. In addition, there were more general themes of consumption patterns, poverty, environmental quality, health, the quality of life for women and children, the international economic environment, and food security.

Obviously, all these themes are important; however, a crucial shortcoming of this type of theme-setting – very typical of conferences of this sort – immediately becomes obvious. If there are 31 focal points then, in reality, there are none. Even political systems have a limited capacity for dealing with problems. If everything is important, nothing is done. The politics of 'muddling through' continues, and genuinely ambitious new strategies peter out.

So what is the right starting point? It is necessary to set strategic priorities to achieve a genuinely innovative movement. Such priorities must not be selected arbitrarily or accidentally, because that would only lead to endless discussions about which is the right one. Any such priority must not only open the door to the solution of one of the many issues, but also provide a pathway to the solution of the global problem. Setting priorities becomes legitimate once there is a clear ranking of dangers.

Further consideration of Agenda 21 therefore shows that almost every problem area is predominantly an issue of energy supply:

- Climate change is caused by the combustion of fossil fuels and the uncontrolled release of methane from the supply of food energy for humans and animals; by the destruction of forests via acid rain, which is caused by the use of fossil energy; by the destruction of the rainforests, caused principally by the demand for wood fuels and clearance of forest acreage to grow food; and by injecting additional heat from nuclear power plants into the atmosphere.

- Destruction of the ozone layer is caused primarily by the use of CFCs for refrigerators or air-conditioning equipment, but also by air transport using fossil fuels.

- Cross-border air pollution occurs through the use of fossil energy.

- Soil erosion and water pollution are in large measure the result of highly concentrated use of chemical energy and concentrated organic waste.

- Desertification is mostly the result of abusive clearcutting of vegetation for essential fuel where there are no other energy sources.

- The environmental damage caused by shipping is due primarily to energy wastes, as is the destruction of marine life.

- The population explosion is deeply rooted in the fact that people – because they have no access to usable energy – seek survival through the classic escape route of increasing human work energy by increasing the size of their families.

- Poverty is prevalent mostly in those regions of the world without sufficient energy sources and services, the basis of all economic development.

- The fact that environmental quality and human health are endangered by current energy supply patterns needs no emphasis in the age of smog in urban areas and of Chernobyl.

Man's economic activities are at the centre of societal development. In turn the core of any economic activity is the use of readily available energy. Those who regard 'energy policy' or the 'energy industry' as one of several policies or economic sectors – as a single, speciality issue – have neither recognized nor understood the basic cause of the dynamics of environmental destruction. If they do not understand that, they cannot find a solution. It is no accident that the history of human development has always been a history of the various energy supply options. Environment, agriculture, industrial production, transportation, development – these and other sectors of the present age, which thinks and acts almost exclusively in terms of division of labour, are primarily sectors of energy politics and the energy economy. To a vastly underestimated degree, they affect international, national and local politics and policies and their structures. Questions of power or dependence, wealth or poverty, privilege or equality, destruction or survival of human societies have always been decided by that key criterion of who has access to energy. This has now evolved into a question of the survival or death of global society.

Energy is the lifeline of any natural and societal development. No natural and no societal life process can be imagined or described without reference to its energy requirement. Ecology deals with issues of energy conversion and the transformation of matter in ecosystems. For this reason, the manner in which we consume energy has created more than a narrow set of problems that can be solved by a single industry sector or a specialized political department. All humanity is threatened with decline because the energy supply system of the 20th century has all-encompassing destructive consequences – the exhaustion of oil, coal, natural gas, nuclear energy, and the self-destructive energy exploitation of developing countries, including loss of their vegetation. All this has produced a cancer in society's organism that keeps metastasizing in more and more parts of its body and is systematically eating it away. The energy crisis becomes a crisis for humanity.

The significance of energy supply as the heart and circulatory system of all life cannot be changed. What can be changed is the choice of our sources of energy. Humanity has the chance to survive only if it is able, within a short period of time, to replace conventional energy sources with the solar energy that flows through the planet's ecological system. Ignoring these pervading opportunities so thoroughly and for so long has created a state of affairs where their continued neglect has assumed the dimensions of a global threat to the common good. Wilhelm Ostwald, a Nobel Prize winner in chemistry, in his 1909 book on issues of natural science and philosophy, *Energetische Grundlagen der Kulturwissenschaft* (Energetic Foundations of the Cultural Sciences), pointed out, for example, that the sun with its radiation constantly sends us 'free energy', and that this process drives 'practically everything that happens on earth'. Ostwald had already distinguished between two forms of solar energy. On the one hand there are the daily 'newly captured and converted radiation energies which, in economic terms, represent *regular income* and which may be *consumed* on a regular basis, after deduction of the required reserves'. On the other hand, there are

> capitalized stockpiles in the form of fossil fuels . . .
> We are dealing, therefore, with a part of our energy
> system that behaves in some way like an unexpected
> inheritance, persuading the inheritor temporarily to
> lose sight of the principles of a lasting economy and
> to live for the day. It must be emphasized here that
> even thrifty consumption merely postpones
> exhaustion, but does not prevent it.

His conclusion, drawn long before our knowledge of disastrous environmental dangers, was: 'An enduring economy must be based exclusively on the regular utilization of the annual solar radiation energy'.[5]

It is not enough to increase the share of renewable solar energy to 10, 20, 30, 40 or 50% of human energy consumption. It would not overcome dangers to our existence, but would merely postpone the collapse of human civilization. *The goal for the century ahead must be the complete substitution of conventional sources of energy by constantly available solar energy – in other words, a complete solar energy supply for mankind.* This is the only way to achieve the permanent recovery of mankind. It is the essence of the most important law of natural science, the Second Law of Thermodynamics, which addresses the issue of entropy, and which has been described by the German environmental journalist Christian Schütze as the 'Basic Law of Decline', a decline that can be stayed only with the help of solar energy:

> We can turn and twist as much as we want, but our problems will remain unsolved, as long as we are unable to tap the only real source of negative entropy so intensively that it covers our entire energy demand, including the energy needed to keep material entropy as low as possible.[6]

That this is possible, and how, will be shown here. This book was written as a contribution to the goal of turning the theoretically possible utilization of solar energy for all civil energy needs into a real possibility. To that end, the basic condition is to make the conversion of mankind's energy supply system to solar energy the very top strategic goal of politics and economics – *Priority Number One.* This claim will be regarded as excessive only by those who still underestimate the all-encompassing, disastrous consequences of our current energy sources, and who fail to recognize the central importance of energy supply. If we really want to succeed in tackling the problems outlined in Agenda 21 we have to arrive at 'Agenda 1' – the all-encompassing solar energy economy.

Solar energy is not a nostalgic remembrance of times past, of an earlier pastoral idyll, which was anything but an idyll for most human beings. Rather, the conversion of man's energy system to a solar one, and with it the reintegration of humanity's energy system into the planet's, will be the decisive step towards incorporating humanity into 'the rhythm of nature'.[7] This opens up new opportunities in almost all relevant fields of action where we are now running out of options, and which those in charge refuse to see in an energy context. This all-encompassing utilization of solar energy will offer a new perspective for humanity – a realizable, concrete vision, a realistic utopia. This project embraces far more than replaces current energy supply technologies with solar energy technologies. This fundamental shift is meeting with such huge obstacles and opposition because it challenges the very web and interplay of energy, economics and social systems.

The introduction of a global solar energy system has more fundamental importance for humanity than the Industrial Revolution and the French Revolution had for the economic and political development of modern times. Both revolutions led humanity into a new epoch, but one that no longer assures permanence. Only all-embracing solar energy utilization promises the kind of development that will benefit humanity permanently, because it makes possible universal human rights and the right to self-determination in all societies. Only a global solar energy system permits an ecological economy, a humanization of the Industrial Revolution and the transfer of both to all people. The Industrial Revolution's path has so far led mankind into deeper schisms and disruption. The reversal of social achievements that threatens now, even in the traditional centres of the Industrial Revolution, will lead to rapid self-destruction. A solar energy system, on the other hand, opens up unique and – because of the basic importance of energy – manifold opportunities, as opposed to the multiple dangers caused by the existing energy system.

Unreal Realism

The central objection is well known: conversion of mankind's *total* energy supply to solar energy is considered unrealistic. This can mean only two things: either that man's self-destruction is unavoidable, or that there are other, more realistic prospects for the survival of mankind. No other possibilities exist.

The plea for solar energy is not a technological one, but a political one. Nor is it another analysis of the dangers of destroying the environment; it deals with the inadequate political efforts to avert crises and the deplorable lack of perspective. It both aims at, and culminates in, a strategic design. Politics, in the original meaning of the word, is the forming of society according to values that are potentially valid for all people. In Greek philosophy, political action meant not only action on behalf of the community – in contrast to private, self-oriented and self-serving action – but also a special way of shaping society: a form of action that takes its cue from ideas of equality and freedom, rather than despotism and tyranny.[8] This kind of rethinking should be at the core of political discussions at a time when just about every activity is labelled 'political', even if there is no positive reference to the public at large. Terms such as 'special-interest politics', 'company politics' and 'union politics' are all, if interpreted rigorously, inherent contradictions. The degradation of the very concept of politics goes hand in hand with the increasing failure of politics to come to grips with the requirements of a humane shaping of society.

An intrinsic part of this goal is to protect society from life's dangers. For this reason, state security has always had a leading role, not subject to much controversy even in many democratic societies. Now, for the first time in history, the question of maintaining the natural foundations – and with them the social foundations – of life has become *the* security matter for the future, even for those societies not yet subject to frequent natural catastrophes. For the first time, classical foreign and security policies are unable to combat the principal threats. 'What use is the best social policy when the Cossacks are at the gates?' German chancellor Adenauer used to ask in the 1950s to justify the primacy of military strength in Western politics. Today we have to ask: what use is military security if the brush fire of our energy consumption burns, dries up, sweeps away and irradiates everything that lives or is needed for life? In 1961, the German political philosopher Dolf Sternberger described the maintenance of peace as the principal subject of politics, because it is the basis of any society's life.[9] This is even more fundamentally true of peace with nature. The maintenance of the natural foundations of life must be the 'People's Road' of politics.

In the face of this challenge, those who claim to be practising *realpolitik* are moving on a shaky, unreal foundation. Not only do the governing paradigms of politics and economics deprive large segments of the human population of life's fundamentals, but they also render those available in developed societies increasingly uncertain. Those mainly responsible for this state of affairs cannot even come up with an appropriate counter-concept, and that is why their policies must fail. A 'realism' incapable of warding off the essential threats to society's structure can no longer be called 'political'. At best it is a selfish realism, fighting for a place for oneself among the elite who, just possibly, may have the dubious privilege of being the last to be defeated. Political action has atrophied to the point where it is merely the best means of coping with existing structures. It fails as soon as a political answer to real problems demands questioning these structures and moving beyond them. Oskar Negt and Alexander Kluge, observing these 'relational measures of politics', have argued that *realpolitik* turns out to be 'useless' if one ceases to be fascinated by 'the artistes of the possible' but looks instead at 'the visible results of this type of 20th century politics'. It

> does not produce anything of permanence and, for that very reason, no (sense of) community . . . Faced with interests that took their cue from (a sense of) community and that saw themselves as political forces, *realpolitik* has always invoked the derogatory point of view of the merely utopian and thus contributed to the mystification of the real power of the status quo. What we are criticizing in this realpolitik is not its momentum of realism but that it is imaginary, without reality.[10]

Whether a new political design, in the sense of the original meaning of politics, is realistic depends on two criteria:

- whether such a strategic concept can provide a *consistent and non-contradictory solution* to the real threats to the general public;

- whether there is *freedom of choice* to grasp the necessary options to act in practical terms – in other words, whether real alternatives for action are possible.

For a solar strategy this means that if there were insufficient solar energy sources and no technologies suitable for their utilization, or if they were to cost more than the available financial means, there would not be, despite all perspectives implied, the freedom to take this route. If, however, a concept meets both criteria, then it is realistic.

1

The Industrial Revolution: Turning Away from the Sun

Energy statistics generally cover the consumption of commercially produced primary energies such as oil, coal, natural gas and biomass, and their utilization as fuel for the production of electricity, with nuclear power listed separately. 'Non-commercial' energy forms are sometimes listed as well – vegetation-based fuels used directly without a commercial detour as long as the supply lasts and is within reach. However, all these statistics are incomplete and, at the same time, symptomatic of the limited understanding of energy. They do not take into account, for example, the passive use of solar energy – in other words, the use of solar energy without reliance on technologies, which in the summer, or in warm climates, renders heating energy unnecessary. Nor do they cover the most essential and most primary energy sources and energy converters: food for humans and their domestic animals, as well as their ability to convert biomass into energy, which are therefore also forms of solar energy. Human energy usage and consumption and, with it, the solar share are larger than indicated in any energy statistics.

The exclusion of food from energy statistics reflects two factors: that mankind sees itself only as a user and no longer as an integral part of the energy system, and that the energy system is no longer regarded as an integral part of nature's cycles. This shows an alienation from the elementary foundations of life as well as a lack of historical perspective in contemplating energy supply systems. For these reasons, humans generally are no longer conscious of the significance of energy systems for the development of human society as such. The times are long past when human labour and technical energy utilization were regarded in a sociological and energetic context, a mode of observation exemplified, for example, by Serge Podolynski in 1883 in his essay 'Human Labour and Unity of Power'.[11] Those ideas, which are being rediscovered by science today under the generic term of 'Ecological Economics', represent a dimension that has been ignored, forgotten or neglected by the science of economics in the course of the 20th century, but which is nevertheless the most significant one affecting economic activity. Essentially, it is a way of contemplating the energy flow and its short-, medium- and long-term economic, social and ecological consequences.[12]

Energy Conversion and the Transformation of Civilization

Earth is, in its totality, an energy conversion system. Jean-Claude Debeir, Jean-Paul Deléage and Daniel Hémery have clarified those linkages, central to our future and of which we have to be aware today more than ever before, in their book *Energy and Civilization through the Ages*.[13] Among living energy converters they differentiate between 'autotrophic organisms' (plants that are able to store solar energy in their structure as chemical energy after conversion, typically green plants) and 'heterotrophic organisms' (man and animals that feed on plants and other heterotrophs). Mankind develops methods for more effective energy generation, from plant cultivation to agriculture, from raising livestock animals to the use of 'artificial converters' such as tools and machinery. Energy consumption of plants and animals is limited by the volume and size of their bodies. Man, by being able to create aids for energy utilization, takes steps that lead to 'domination of the earth'.

For a long period mankind itself was the most efficient energy converter by means of its own muscle power, and this is still the case today in many regions of the globe, when no other energy source or technology is available. Even today, people are still the most versatile such converter. By developing tools, they were able to create mechanical energy using their bodies. They hunted and gathered as far as their feet would carry them, then later shifted to agriculture and animal husbandry, which necessitated the formation of societal systems of human order. Two features characterize this development: first, the actual influx of solar energy continued as primary energy; second, human settlements could only spread as far, and their population could only grow as large, as local energy potential permitted.

The next stage consisted of the development of chemically bound 'inactive energy', beginning with the discovery of fire and the possibility of burning wood, making available for the first time fairly large amounts of energy. Men realized the value of this energy conversion to save labour or to perform work that it was impossible to achieve with humans or animals as energy converters. These early phases of the history of civilization show that they are already closely linked to the energy supply structure, without which no further societal development is possible. These phases are long: it was about half a million years ago when man first managed to keep alight the fire started by a bolt of lightning. The first tools came into use about 35,000 years ago. Later, between 4000 and 3500 B.C., after the first sailing ships and windmills were developed and the use of hydropower began via dams, irrigation systems and machinery using rivers and creeks, cultural development began to accelerate.

The first centres of civilization arose in places where irrigation techniques were used. Food production – plants as 'biological solar collectors' – was increased, requiring an increase in the ability to organize society and, correspondingly, political stability. The population could grow. Sailing ships permitted long-distance transport of foodstuffs at low cost, as did wheeled vehicles and the use of draught animals. A phenomenon began to develop that has increasingly characterized civilization to this day: transport infrastructures and means of transport make people independent of the food supply produced in the immediate vicinity; energy production and energy consumption become uncoupled.

As human labour was still the most versatile type of energy converter, acquisition of slaves began to expand over an increasingly wide area, resulting in the establishment of corresponding hierarchical forms of culture and motivating warlike expansion. This kind of energy service controlled the rise and fall of societal systems and political power relationships for a long time. As long as slaves were cheaper than energy conversion techniques, they formed the foundation of the supply of labour and food energies. Seen against this background, the slave rebellion led by Spartacus rattled the very foundations of the entire political and economic order of the day.

Because of unrestrained clearing of forests and soil erosion, many humans lost the basis of their existence, triggering mass migrations. With the invention of windmills and watermills – in different societies at different times – it became possible during the past 2000 years to perfect irrigation techniques for the production of food energy, including grain processing. Smelters were built for iron production, causing excessive forest exploitation; water-powered foundries then improved productivity. Coal deposits were discovered and used for iron smelting or for heating, but this was initially hampered – as was the use of the forests – by the difficulty of transporting bulky fuels.

These individual development phases were not identical in different regions of the world, though, in terms of the chronology and discovery of conversion techniques. There are many examples of how some societies had an advantage in energy technology – and thus a general developmental headstart – but remained at a standstill for too long, only to be caught up and surpassed by other societies with even more advanced energy-technological innovations of their own.

The decisive breakthrough that made possible the wealth of the industrialized societies and which, at the same time, explains the shift from earlier regional wars to the present global war against the environment began with the development of the steam engine and, shortly thereafter, with steam navigation. This was the real Industrial Revolution, the basis for the rise of industrial societies. Until then, industrialization had been held back by two bottlenecks: the inability to transform heat into motion, and the limited means of energy transport. By overcoming these restrictions, energy consumption and industrial development accelerated rapidly. This occurred in stages, from the effective exploitation of coal deposits to oil and natural gas fields on a global distribution scale; from the development of steam-powered shipping to railways, cars and trucks and finally aeroplanes; from the use of coal tenders (for railway locomotives) to the construction of oil and gas pipelines, and the introduction of oil tankers. It became possible to transport primary energy anywhere. Then, with the advent of electric power generation and the construction of electricity grids, the energy supply system became even more flexible, and power transport accelerated further. Finally there was nuclear power. The Industrial Revolution was a revolution of energy technology; it launched the apparently final abandonment of direct and indirect use of existing solar energy.

Human and animal muscle power were replaced by modern energy conversion techniques where it was cheaper for users and operators, where primary energy was readily available, and where capital could be utilized. The accumulation of capital and, with it, the creation of large metropolitan industrial regions began around traditional energy centres and, later, near refineries, shipping and rail lines, pipelines and long-distance power transmission grids. These developments caused labour shortages and increases in real income, and launched in turn mass migrations to these centres, resulting in the creation of cities with millions of residents and causing misery and poverty in rural regions. The increased productivity of labour, which primarily increased energy productivity, also increased the surplus per worker and thus contributed substantially to the creation of economic class strata.[14]

Energy efficiency was further increased when social movements occurred in the wake of the Industrial Revolution. The successful fight for shorter working hours, higher wages and better working conditions increased costs and contributed to the accelerated replacement of human labour by mechanical energy converters. The constant technical improvement of energy conversion techniques continuously reduced the cost of energy and in turn accelerated industrial growth and the consumption of raw materials and energy in a steeply rising curve. Trends in productivity, which in the increasingly internationalized economy became the prime factor determining the stability of political systems, were characterized mostly by increases in energy efficiency. We can only understand the full significance of this if we consider not only 'exosomatic agents' – tools and machinery – and their energy requirements, but also the continuing process of replacing human labour energy by electronic data processing, computer-assisted production and communications technologies. Time and again short-sighted economic interest groups who persist in maintaining the status quo fight such developments because they do not immediately recognize the medium-term advantages for the national economy.

Energy Supply and Migration Movements

Generally, the Industrial Revolution and population growth can be equated. But the reason for population growth in industrial centres had more to do with migrations from the country to the city and with international migrations to the centres of mass production and labour than with the growth of the native population. These movements also masked statistically another sociological process that is just as fundamental: having secure and cost-efficient energy sources and technologies at one's disposal gradually led to a stagnation in population growth! While it is difficult to unearth the link between meeting the energy demand and demographic trends in the many analyses of the population explosion, it is statistically obvious and is the inevitable conclusion to be drawn from experiences with the social effects of energy systems. Table 1, derived from the *World Energy Atlas*, indicates clearly that in the countries with the highest energy consumption population growth has stopped – demonstrated clearly in the European Union and Northern Europe.[15] Only Canada, the United States and Australia register an annual population growth of 0.7–1% despite their energy saturation, but this has to do with their immigration figures. The statistics from Latin America, Asia and Africa show that

Table 1 *Energy consumption and population growth*

	Energy consumption per capita 1985 (ton of oil equivalent)	Annual population growth 1985–2020 (%)
Canada	9.0	0.8
United States	7.5	0.7
Northern Europe	5.4	0.2
Australia, New Zealand	4.8	1.0
Former Soviet Union	4.7	1.0
European Union	4.0	0.1
Japan	3.0	0.7
Latin America	1.0	2.2
Asia	0.4	2.4
Africa	0.4	3.0

the greater the energy shortage, the more pronounced the population growth!

But even if we take into account other development indicators (life expectancy, national product, agricultural production, water supply, literacy rates) we stumble time and again over the essential connection between the availability of energy and a society's development capability. Table 2 on page 20 shows unequivocally that a high rate of energy use is synonymous with population stagnation and high national product. Lack of energy is linked to population explosions (even with low life expectancy), poor national product, a high share of agricultural production and a rural population, and poor water supply.[16] To draw the conclusion that 'the future belongs to the city', as did the authors of the 1992 report *State of the World Population*, published by the United Nation's Population Fund, is a pointless reaction. If Sao Paulo grows from 7 to 25 million inhabitants between 1980 and 2000, Buenos Aires from 10 to 25 million, Mexico City from 9 to 30 million, Calcutta from 9 to 20 million and Bombay from 8 to 20 million, and if the slums, avalanches of mud and the danger of epidemics grow, then the answer must be sought in preventing the flight from the countryside. Just as futile are the innumerable, usually almost identical, recommendations to slow population growth, typically via literacy, education and training programmes or in programmes for family planning and strengthening the rights of women who in some regions spend 300 days of the year collecting firewood.

As long as there is no energy available to replace human energy, it will be impossible to slow population growth. At the same time, these migratory movements have their origin mainly in rural areas of developing countries because their growing populations see no prospects for economic development. Without lasting and available energy agricultural production cannot be increased, nor can the lack of water be remedied, nor can crafts and small industries flourish, nor can vegetation be saved.

Only China has been able to bring its population growth under control and prevent flight from the land, despite a level of commercially available energy that is lower than that of other medium-developed countries. In addition to heavy political pressure, the main reason for this has been China's

Table 2 *Comparison of energy, consumption, economic and living conditions*

	Commercial energy consumption per capita (kg oil equival., 1990)	Expected pop. rate (1990–2000)	Average life expectancy at birth (1990)	GNP per capita (US$, 1989)	Agric. production (as % of GDP)	Rural population (as % of pop.)	Population with access to safe water (as % of pop.)	Adult illiteracy (as % of pop.)
Industrial countries	4,930	0.5	74.5	17,017			100	
Medium human development								
development	1,061	1.6	66.2	1,998	13	42	85	21
China	591	1.3	70.1	350	32	67	74	27
Low human development								
development	160	2.7	55.2	384	28	71	46	46
India	226	2.0	59.1	340	30	73	75	52
Least human development								
development	63	3.0	51	237	40	78	46	55

successful agricultural reform compared with other developing countries. Instead of trying to convert agriculture to the widespread use of machinery and fertilizer along the lines practised in industrial countries – for which most farmers lacked the capital – the country's leadership organized a cost-effective cycle of using biomass energy with corresponding low-tech converters such as biogas plants.

Where people cannot buy energy and cannot use manpower-saving energy converters, family members are the most readily available substitute. In other words, as long as the lack of energy in developing countries remains a fact, literacy programmes, birth control and women's rights programmes will do little to change the patterns of population growth. That this key point is not mentioned even in the margins of so many analyses of the population explosion is explained simply by the fact that even the world of science knows nothing about the socio-economic importance of energy systems, and the human factor is ignored in energy analyses. We may measure the power of cars in horsepower, but hardly anybody then draws conclusions for the analysis of sociological structures where such engine power does not exist. The same holds true for human muscle power. Energy consumption in the industrialized countries is a replacement for servants or slaves: the average per capita energy services available in the industrial countries correspond to the use of more than one hundred slaves. Where those energy services are not available, people resort to no-cost human muscle power as far as possible, or nature. Figure 1 on page 22 shows energy consumption in primitive cultures, hunting cultures, early agricultural societies, developed agricultural societies, early industrial and modern technological cultures. It confirms what was already evident about the use of energy and technology in agriculture: industrial-technical development corresponds to a drastic increase in energy consumption.[17]

While biomass accounts for only 3% of the energy supply share in industrial countries, it is 35% in developing countries. In Ethiopia, Burkina Faso, Tanzania, Uganda and Nepal, energy consumption based on biomass is more than 90%.[18] At the same time, biomass is used more and more in non-renewable ways, making the decline in its availability obvious in the foreseeable future. Without any new energy supply, there will be an even more

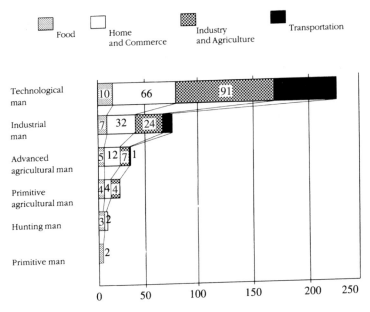

Figure 1 *Daily energy consumption at various stages of human culture (in 1000 kcal)*

Source: Earl Cook, 'The flow of energy in an industrial society', *Scientific American*, no. 9/1971, p. 136.

rapid population increase and a further increase in migration movements. But for economic reasons, the type of energy supply offered by the current global energy system remains unaffordable and unreachable for developing countries in the amounts needed for their economic development. The kind of change that will be required is obstructed by a fully established 'political-economic energy complex' that totally overshadows the well-known political influence of the 'military-industrial complex' – a political powerhouse that has tailored control of energy sources and their utilization to the needs of the industrialized countries.

Energy Imperialism

For centuries, the European powers have pursued colonial imperialism to ensure their control over energy sources and other raw materials by military force. The colonial period has ended and been replaced by an energy imperialism. This is economically more advantageous for the industrialized world because, under normal circumstances, it can forgo direct political repression of other countries and can operate without carrying the burden of responsibility for their administration.

More efficient energy conversion increases the economic surplus compared with those who are technically backward. Greater energy income increases the room for economic manoeuvring, permitting the exploitation of the cheapest available energy resources anywhere on the planet and employing even more energy technology. An international energy market is created through transport opportunities. Expensive energy is replaced by cheaper energy, a situation that always leads to harsh collisions of divergent interests.

The international history of energy can be retraced only by taking into account the interwoven political, institutional, economic and social factors. Market prices were regularly manipulated by governments and companies offering energy, and frequently have little to do with actual costs.[19] Industrial countries fought over coal mining areas, as exemplified by the Franco-German conflict over the Ruhr area following World War I. Coal producers defended themselves against the inroads of the oil suppliers. These massive conflicts between industrialized countries and companies are reduced as crude oil increasingly forces its way into energy markets, and the oil resources of non-industrialized or colonized regions of the world are discovered and exploited economically.

Oil, the energy carrier with more uses than any other and the cheapest to produce – because of the simple, yet decisive advantage of being a liquid – has had the central role in the world's energy supply since the beginning of the 20th century. The early oil concerns, able swiftly to gain a profit advantage over other energy companies, focused their attention on the acquisition of production rights in foreign territories, secured politically by their governments. In the world energy order of the 20th century, domestic energy resources are no longer the decisive factor for Western industrial countries; instead it is the ability to employ capital to control access to energy sources – including the advantage of sparing their own by exploiting foreign resources. 'Persian oil is yours', President Roosevelt told the British Ambassador in Washington, Lord Halifax, in

1944, as reported in Daniel Yergin's Pulitzer Prize winning book, *The Prize*: 'We share the oil of Iraq and Kuwait. As for Saudi Arabian oil, it's ours.'[20] Roosevelt's words characterized modern energy imperialism, in effect a division of labour initially among the oil concerns and their governments and shared later by the oil-producing countries.

The discovery and exploitation of the oil wells of the Middle East brought together oil concerns and, later, Western governments as well. As early as 1928 they formed the Achnacarry Cartel, which – independently of the source of the oil

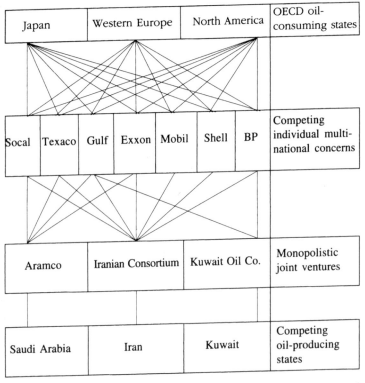

Figure 2 *Structural links among the oil-producing states, the multinational oil concerns and the oil-consuming nations in the world market before the 'energy crisis'*

Source: M. Massarat, *World Energy Production and Reform of the World Economy*, Frankfurt/New York, 1980; p. 169.

– set prices, stabilized and delineated markets, and set up joint planning to expand distribution networks.[21] The colonial powers were replaced by companies with mineral and oil exploitation rights in the Middle East that at first did not refer to quantities of oil to be produced, but to exploitable territories. Thus the profit shares of the various countries became dependent on the amount of oil produced. Only in the early 1950s did the oil multinationals agree to split their revenues with the producer countries on a 50/50 basis. Figure 2 shows which channels carried the oil flood into the industrial centres of the United States, Western Europe and Japan until the early 1970s, mostly via the 'Seven Sisters'.[22] Their power and market position were only minimally affected by the establishment of state oil companies in Italy and France, for example, with contacts of their own with producer countries, although these state companies had been set up with the express purpose of making those countries independent of the leading multinationals. The attempt of Iran's reformist Prime Minister Mossadeqh to gain control of the production rights by nationalizing the country's oil wells ended with the international boycott of Iranian oil by a coalition of oil multinationals with the British and American governments. Iran's production dropped 90% by the time of Mossadeqh's overthrow engineered by the CIA (1953).[23]

A high level of energy supply based on dumping prices became the lifeblood of Western industrial societies, which use it to secure an increasingly large advantage over the rest of the world – in production, in consumption and, via taxes on crude oil, in their government revenues. Only at the beginning of the 1970s did the oil-producing countries succeed in gaining user rights for their oil wells. The collaboration of the oil producers in OPEC (Organization of Petroleum Exporting Countries) and the existence of the Soviet Union superpower with influence in the Middle East made it impossible to launch sanctions of the type used earlier against Iran. The oil crises of the 1970s were the manifestation of the justified desire of the producer countries to obtain fairer prices for their products and to eke out their own resources. The economic growth and the social services offered by the Western democracies after World War II were based mainly on cornering the oil belonging to others at prices that reflected neither the economic nor the political truth – let alone environmental truth.[24]

However, for a number of reasons, OPEC does not endure. The Iran–Iraq war of the 1980s created a demand for additional money that was met by a one-sided increase in oil production. Saudi Arabia, Kuwait and the United Arab Emirates obtained the political support of the West as a hedge against internal revolutions, propping up feudalism in the name of the Free World. Because of their voluminous investments in the industrialized West, the oil states lost their interest in driving the West into economic crisis with new, drastic price increases, because that would endanger revenues from these investments. 'Oil needed markets, and markets needed oil; the calculation of mutual self-interest should be the basis of a stable, constructive, non-confrontational relationship that would extend into the twenty-first century.'[25]

In other words, the 'political-economic energy complex' established itself after the oil crises of the 1970s and early 1980s, after the oil-producing countries had insisted on being included. Libya, a political exception to this well-oiled harmony, is for that very reason a continuing thorn in the side of the West. When Iraq threatened to become another impudent exception with its attempt to take over Kuwait, the largest military expedition since World War II was launched in 1991, made politically easier by the collapse in 1990 of the Soviet Union as a competing empire. Two years earlier, Iraq had committed a much more serious violation of international law than the occupation of Kuwait – that is, the use of chemical weapons – but there was no reaction by the international community because the question of political control of oil was not at issue.

Not only does energy imperialism evidently determine the rate of growth of the world economy, but it also exerts a crucial influence against efforts towards environmental repair and liberation from dependency in energy supply relationships. The catastrophic debts of the developing countries are due mainly to the oil crises. These countries owed a total of US$64 billion in 1970, but by 1982 their debt had risen to $830 billion and to $1,265 trillion by 1990. For oil purchases alone, these countries had to spend an extra $260 billion between 1973 and 1981 directly, without taking into account any indirect consequences.[26] The second oil crisis dampened the global outlook abruptly, raised indebtedness and interest rates in the industrial countries, and between 1979 and 1982 it alone caused a 40% increase of foreign debts among developing countries. The Western industrial countries and the oil-producing countries were the winners in this crisis, to the extent that their common interests became even more closely knit.

The debt crisis triggered by the oil crisis drove the developing countries to increase their production of mineral raw materials to make up for losses in income – and led to a collapse of prices that in turn added to the debt mountain. The result today is that there is no scarcity of energy and raw materials, but rather an excess of supply – despite the fact that both are finite and exhaustible. Because the industrial countries have, at the same time, far better financial, technical, industrial and administrative means to save resources by efficiency improvements and substitutions, they exert more leverage, which they employ, in concert with the oil-producing countries, against the developing countries. Part of this strategy is to maintain developing countries' dependence on energy, and thus their political dependency. Energy prices, with the oil price as an index, are kept at a level low enough for the short-term competitive goals of the industrial countries, but this level suppresses developing countries. Compared with the international benchmark price, gasoline and diesel fuel prices in Uganda in 1985 were higher, for example, than in the Federal Republic of Germany, according to a study by the World Resources Institute, and in India, Ethiopia and Kenya they were higher than in the United States.[27] It is clearly shown by Ian Smart in his analysis 'Energy and the Power of Nations' that the energy demand of developing nations permanently increases the international power of the oil exporters.[28]

At the same time, every effort is made to conserve and expand the energy supply structures that create and sustain these dependencies. Even the efforts of the leading Western development banks (the World Bank, the International Monetary Fund, and the Inter-American Development Bank) as well as the development funds of the oil-producing countries have had only a small, or even counterproductive, effect on developing countries: by the construction of highways, for example, they encourage motor traffic and fuel consumption. The largest share of development aid from all countries for energy development – considerably more than 90% – is concentrated on non-solar energies.[29] As long as the developing countries rely on such aid payments, they remain dependent on industrial and oil-producing countries, which will never offer energy at prices corresponding to their purchasing power: the very existence of the pricing system, marketing structures, the industrialized countries' own energy needs, limited resources and global environmental burdens stand in the way.

The North–South distribution battle is moving further and further away from exploitation and control rights for energy and raw materials towards rights for quotas for environmental burdens. Within the new frame of reference of ecological reasoning, the industrial countries are now attempting to limit the energy consumption of developing countries. But the only adequate conclusion – concentrated efforts to foster solar energy use – is not open to debate because it is at variance with the influences and interests of the world energy system. The leaders of the international energy system will neither permit others to share in increasingly scarce resources, nor take any initiatives of their own to construct an alternative energy system that lies outside their control.

A comparable picture emerges in the structure of an international cartel of electrical equipment producers that has taken shape since the beginning of the century: domestic and, later, international markets were divided up by means of marketing agreements and with the help of major commercial banks – in Germany, between AEG and Siemens together with Deutsche Bank (whose first head, incidentally, bore the name of Siemens as well); in the United States between General Electric and Westinghouse. 'Home protection agreements' were signed, and 'cross-licensing', the pooling of patents for common utilization, was practised. In 1930, the International Notification and Compensation Agreement (INCA) was signed, followed in 1936 by the formation of the International Electrical Association, which tried to eliminate unwanted competition via its 'factory committees', and which established committees to fight off outsiders. In Brazil in the 1960s, for example, the indigenous electrical industries were squeezed out, with the help of the bribed Brazilian government, by allowing the tax- and duty-free import of electrical equipment from American producers while domestic manufacturers still had to pay taxes. The system carefully distinguishes between 'producing' and 'non-producing' countries; the latter are to be considered merely as buyers or locations for domestically produced cheap products, but not as independent manufacturers of electrical technology and power plants.

In a widely publicized 1963 trial in Philadelphia, directors of both General Electric and Westinghouse were found guilty of price fixing and market allocations, leading Chief Judge Ganey to comment:

> This is a shocking indictment of a vast section of our economy, for what is really at stake here is the survival of the kind of economy under which this country has grown great, the free enterprise system.

Ganey also agreed with the characterization of the defendants by Assistant Attorney General Bicks as having 'flagrantly mocked the image of that economic system of free enterprise which we profess to the country', with the court adding, 'it has destroyed the model, it seems to me, which we offer today, as a free world alternative to state control, to socialism, and eventual dictatorship'.

Organized along the patterns of a secret service, the web extended, according to a 1973 book, *The Electrical Equipment Conspiracies*,[30] as far as Japan, whose corporations became, beginning in the 1960s, party to these agreements. These cartel agreements are categorized by product sections – from domestic current meters to electric motors and power plant turbines – and by market territories. No wonder that these 'electrical equipment conspiracies' will seek to obstruct the production of solar energy technology as long as they do not see the time as ripe to move into these new products themselves. And since the market leaders are identical with the suppliers of nuclear technology, they have a clearly defined interest not to become their own competitors.

The bag of tricks of the energy imperialists includes the targeted exploitation of the dependency of others, manipulation of international energy prices, the occasional granting of development aid handouts to exploited countries, the special support and corruption of loyal supplier governments and, increasingly, military security measures. This becomes most evident in the new strategy agreed upon at the NATO Summit in

Rome in November 1991, which postulated the formation of rapid deployment forces against 'threats from the South' – 'threats' meaning the hugely inferior, in military terms, and politically divided Islamic states.

The energy revenues of the world energy system's profiteers increase the contrast between the rich and the poor, developed and less developed countries in increasingly extreme fashion, and they far overshadow the colonial exploitation of the past. The more barbaric the conditions – man against man, man against nature – in the economically and environmentally exploited world of the developing countries, the more the Western industrial high cultures feel morally superior to their victims. Energy imperialism is the most effective of all imperialisms because it is the most subtle type of imperialism, and the most difficult to unmask.

The structure of the world's energy policies is by no means only of historical significance. It is an increasingly relevant and potentially explosive issue of our day. This is demonstrated by the increasingly bloody conflicts over oil – oil, which is, of course, the most heavily used energy source and yet the one that will be exhausted the soonest. Not long ago Peru tried to annex Ecuador's oilfields by military force. The Chechen war was a war between various Russian and Chechen political factions, each supported by parts of the state apparatus, over control of the oil business. Afghanistan, Azerbaijan or Tajikistan are simmering with strife because of the need to provide oil for the industrialized world. The USA's support until quite recently for the most medieval Islamic extremism in Afghanistan reeks of oil. So does the possibility that Turkey, which is a transit country for oil from the Caucasus, would be kept in NATO even if there were a fundamentalist upheaval with persecution of the Kurds on a scale reminiscent of the pogroms against the Armenians. *Blood for Oil*[31] is the extremely apt title of the book by Hans Kronberger, in which he traces the bloody map of world conflict over non-renewable energy resources. There is an increasingly acute and frequent pattern of conflict, which will persist unless and until oil and other non-renewable energies are replaced by up-to-date solar energy.

The North–South Conflict: An Energy Syndrome

The following rough sketch of the development of energy systems provides a few basic ideas about the energy syndrome of humanity, a prerequisite for any strategic discussion of how to avoid life-threatening changes:

1. The industrialized world does not consume more energy than developing countries because it produces more industrial goods and requires more services. The opposite is true: it produces and consumes more because it commands more efficient energy conversion techniques, which have allowed it to acquire capital, market and growth advantages. Increased efficiency of energy utilization therefore does not automatically mean a reduced need for energy. It is true that increases in productivity make possible a gradual decrease in the need for human labour energy and for a single technology, but industry and humans shift their energy needs into areas of consumption that are not essential for the safeguarding of economic existence and which are, to a very large degree, energy-inefficient.

2. The environmental Fall of Man did not consist so much of industrialization *per se* made possible by energy conversion technologies, but of energy conversion's increasing reliance on fossil and, later, nuclear resources. Their immediate advantage is high energy density – in other words, the possibility of achieving high energy performance from a small volume. This facilitated storage and transport, entrepreneurial centralization of energy supply and rapid growth of energy consumption without regard to the finiteness of energy resources and environmental limits. In the meantime, we have begun to learn that mankind can no longer afford to burn up all its fossil and nuclear energy supplies.

3. At the time when energy was supplied almost exclusively from renewable energy forms and solar energy stored in the medium term, the earth had to support a much smaller number of human beings. The total grew from about 200 million at the time of Jesus Christ's birth to 1 billion by 1900, to 2 billion by 1950 and to more than 5 billion by 1990.

United Nations estimates place the total at about 6 billion by the turn of the millennium and about 8.5 billion by 2025. The availability of energy is the central precondition to secure for these numbers a future with human dignity. But this cannot be the type of energy supply using the production structures of the last 200 years because that will ring the death knell for human civilization.

Mankind must find its way back to the sun, but it cannot move back to pre-industrial conditions. Its only chance rests on a solar-technological energy revolution for the utilization of inexhaustible, always available, solar energy, aided by modern technology.

4. The spatial uncoupling of energy supply and energy consumption made apparently limitless growth possible – as long as only the regional limits to the exploitation and nature's capacity for absorbing damage were known and the existence of global limits was not realized. But it was implicit in this recognition of global limits that economic progress meant the destruction of nature, and the chasm between economy and the environment grew wider and wider.

Mankind's alienation from nature was made easier as those responsible were not forced to see the indirect consequences of increased energy consumption. Countries that can afford to import energy and still earn high energy-related revenues can avoid despoiling their own vegetation. The damaging effects are visited upon their creators only indirectly through changes in the world's atmosphere.

5. The existing global energy systems are among the key causes of the misalignment of the socio-political structures between North and South. Where energy sources and technologies are not available, and where people are especially dependent on human energy for work, there is not only the problem of population growth, but social hierarchies and political oppression dominate as well. If, however, energy is sufficiently available, birth rates decline, social hierarchies can be dismantled, the space for individual freedom can be expanded, and democracy can flourish. It also follows that renewed energy shortages will bring about the opposite political and cultural development trends.

6. Faced with the global influence of the 'political-economic energy complex', the developing countries do not have the slightest chance of creating a price-effective energy supply system appropriate to their economic capabilities. Traditional energy sources are, in principle, not an effective means of overcoming the discrepancy in the respective levels of economic development. The dependence of the South on the power of the international energy complex condemns the developing countries to a permanent puppet status – without the opportunity to develop economically on their own and to achieve political self-determination. Above all, the developing countries are always the first and chief victims of energy crises.

It is striking that the central significance of energy supply is hardly ever noted in the innumerable analyses of the problems of developing countries – although everybody knows about the direct link between oil crises and debt crises. Surely this partial blindness is not simply because many are not aware of the intrinsic significance of energy for economic and socio-cultural development. Apparently the view is widely held that it is impossible to alter the basic global energy system – except in the general area of energy savings, which can reduce emissions but do not change the structures of global energy distribution. For this reason, new concepts of how to tackle the politics of development concentrate on all sorts of suggestions, but frequently they address the energy question only marginally. Even less frequently do they recognize the elemental significance of solar energy for promising, autonomous, economical and, at the same time, environmentally benign development.[32]

This look at the potential of solar energy makes clear its all-encompassing significance: solar energy offers a secure supply of energy that cannot be manipulated or channelled by any energy cartel. Further destructive exploitation of fossil or biological resources and the unholy fight for legal emission quotas, which in the end will only add to the burdens of developing countries, can be avoided; energy sources and energy consumption would be coupled again, creating the prospect that novel, decentralized industrial and trade structures can be created in developing countries. Just as importantly, solar energy offers probably the best opportunity – the only one not tried so far – for a 'reduction of population growth, improved living conditions in rural areas, thus reducing the drift from these areas to the cities and also reducing international migration movements', as noted in the Harare Solar Energy Declaration of November 1991.[33]

Energy Strategies and Western Centrism

The West (including Japan) nowadays is the undisputed ruler of the planet. A singular concentration of global power is found here: financial, industrial, technological, military, ideological. The well-known view of Karl Marx, that the ruling opinion is the opinion of those who rule, proves to be true in Western assessments of energy perspectives. Developing countries are permitted increases in their emissions, but their magnitude is determined by measures that the golden West considers barely acceptable or enforceable for itself. Behind all this is the unspoken, but already self-evident, philosophy that global environmental protection and the economic development of others must be permitted only if it does not adversely affect the position of the Western industrial countries.

It is a characteristic of Western self-centredness that it regards itself as the more rational part of the world, more competent to solve these problems, ignoring the fact that it is the prime source of these global problems. It is precisely for this reason that the West does not notice the central contradiction: although energy-saving strategies linked to careful growth rates for developing countries are the dominant theme for global climate policies, those same developing countries are being obliged to convert their economic and social organization to an inherited growth economy that will lead straight to an explosive growth in their energy consumption. If these countries model themselves on the Western example, it is a foregone conclusion that, despite increases in energy efficiency, all intentions of a global environmental policy will wind up as so much waste paper. The inevitable consequence is a much greater use of emission machines, mostly cars. A society based on unconstrained principles of market economics will gulp energy according to the laws of the market. In other words, the opening up of new market economies of the size of China without a simultaneous conversion to solar energy represents an environmental death sentence for human civilization.

It is part of the nature of Western capitalist centrism that it does not want to recognize its destructive effects on others. This becomes easier if these impacts can be dumped onto others – on developing countries or on nature itself. The contaminators of the planet naturally prefer to live in elegant, green, residential areas. There are people who consider themselves conservationists, yet who prefer to obtain their electricity from a distant nuclear power plant rather than from a wind farm if the wind generator disturbs their immediate communion with nature. Hand in hand with that attitude comes a growing attention to environmental problems in their own enclaves, but little attention to much graver environmental problems in the developing countries and for the planet as a whole – even if they themselves are the main culprits. This is the result of the individualistic narrow Western idea that one is interested mostly in those things 'that have a negative impact on our user options': in other words 'only the most immediate and most crass forms of environmental destruction' lead to action in environmental policies, as one German observer noted.[34] One of these contradictions is that the alternatives offered limit themselves to employing merely more efficient technology – avoiding any discussion of the much more fundamental issue of new energy sources.

As, it is claimed, this is the only way of looking rationally at energy, these criteria, as has been said previously, are applied to the totally different situation of developing countries, turning into a kind of 'deadly aid' in the process. A precondition for introducing solar energy in developing countries, the argument goes, is to launch it first in industrial countries, as the developing countries are bound to imitate the industrial countries. There are three basic reasons why it is impossible to transfer the energy system of the industrial North even temporarily to the developing countries of the South.

First, unless they have fossil energy sources of their own, they do not have sufficient foreign exchange to make up for their lack of energy by more imports. Second, they don't have an adequate supply infrastructure to transport energy to those places where it is needed most – and the construction of such infrastructures would take too long and would be too expensive to avoid further collapse into the abyss in time. Third, the environmental burdens would be globally intolerable. Because it is essential that developing countries develop an energy system of their own, they should not attempt to achieve it via the detour of conventional energy sources and technologies but directly with solar energy.

Paul Kennedy, in his widely noted historical study about the rise and fall of the great powers, has described the factors of such a rise: military strength, the gratification of social and economic needs of the people, and the safeguarding of lasting economic growth.[35] If one of these three components of national power fails for a prolonged period of time, decline becomes inevitable. The Soviet Union broke up because it had only military strength left. The United States also bases its global role as leader of the pack mainly on military strength, and it is less and less able to meet the economic and social needs of its population, let alone ensure its economic future. The other two major power centres, Western Europe and Japan, face increasing difficulties. If Kennedy is right, and he predicted the dissolution and national break-up of the Soviet Union in his study written in the mid-1980s, the West with its egocentrism is likely to face unimaginable turbulence and dislocations. The 'First World' exerts such a degree of dominance, including an unprecedented level of uniformity in economic thinking, over the suppressed or declining 'Fourth', 'Third' and 'Second' Worlds that their continued fate depends, for better or worse, on the fate of the West.

Since the West commands a position of superior economic strength, most of world opinion believes that the collapse of the Soviet Union and the end of the Cold War now mean global progress at all levels. But on the question of destroying the environment via energy consumption, this unique hegemony rather threatens the opposite. The Western model is the most destructive, clearly evidenced by the fact that 80% of the world's energy is consumed by 20% of the world's population, despite its higher energy efficiency. Greater growth leads to higher energy consumption in the economy even with the use of more economical energy technologies. More consumption means higher energy requirements. When the West now wants to use its economic prescription for energy reform in developing countries, it acts like an unreformed alcoholic who wants to recommend the proper withdrawal treatments for others. Without a radical shift of the world's energy supply system to non-destructive solar energy sources – without a solar revolution in the wake of the Industrial Revolution – the Western model of democracy and capitalism is not the perfection of history but its execution.

2

Solar Energy:
The Energy Imperative

In order to anticipate the depletion of fossil energies, an intense political effort has been under way since the 1950s to mobilize nature's primary energy forces through both nuclear fission and, in future, nuclear fusion in an attempt to duplicate the sun's energy production on earth. It is evident that nature is being overwhelmed by the continuous combustion of large amounts of fossil fuels. It has become equally evident, meanwhile, that people are overwhelmed by atomic energy. Nuclear energy technology has created a limitless potential for destruction; it is increasingly clear that keeping this potential under control is beyond the power of governments.

It is the tragedy of mankind that it seems almost too late for a counter-strategy. The real dangers of fossil and atomic energy were driven home only after the first catastrophes had already taken place. Before that, so ran the big excuse of our age, it had simply been impossible to recognize these dangers. This then is the Big Lie of the 20th century: the truth is that, for decades, those who could have and should have known, have been ignoring the physical evidence of these dangers. First and foremost, this applies to the scientific elite, who should have had this knowledge before anyone else. The Laws of Thermodynamics have been known for more than one hundred years, but nobody bothered to draw basic conclusions from them that would have resulted in a fundamentally different direction for our energy system. Albert Einstein characterized thermodynamics as the only physical theory that could never be disproved. But for decades energy economics, energy science and energy politics were pursued as though these laws applied only to the functioning of energy technologies in isolation and not to the interplay between technical energy conversion and nature's cycles.

Economic and Natural Entropy

The First Law of Thermodynamics deals with the conservation of energy. Energy can be neither produced nor destroyed, but can only be converted from one form to another. It is present in the most varied ways: as 'resting energy', 'bound energy' or 'unavailable energy'. Energy exists in the form of oil, coal, gas or uranium deposits in closed systems, and does not present a problem as long as it is not released by another energy. It starts as 'free energy', and becomes 'available energy' when it is released and combusted; after its successful conversion, it is 'diffused' and no longer useful energy. The total energy in a closed system always remains constant.

The Second Law of Thermodynamics states that, with every conversion of energy from one form into another, inescapable conversion losses occur. Because of these losses – during the conversion of resting energy into available energy and then into diffused and no longer available energy – heat and other emissions are released that bring disorder to the ecosphere's ordered state, and which in the end bring about its heat death. The term for that is *entropy* – a combination of the Greek words *energeia* (strength) and *tropos* (conversion). Entropy characterizes the transition from ordered systems to disordered systems. There is also negentropy, however, the transition from disorder to new systems of order – perhaps through newly reforested woodlands. The globe, with its atmospheric envelope and its ozone layer, is a relatively closed system. The utilization of bound energy deposits in effect equals the systematic increase in entropy.

When these natural laws were recognized, it was impossible to predict when the increase in entropy would result in a dangerous level of disorder – particularly as it was impossible to predict the breathtaking growth of energy consumption at the beginning of the 20th century. Yet, in 1912, Wilhelm Ostwald had already formulated the 'Energy Imperative' with the words 'don't waste energy, utilize it' – based on 'the general application of the Second Law of Thermodynamics to all actions and events, especially to the totality of human activities'.[36]

Many decades before Georgescu-Roegen postulated, in his classic work *The Entropy Law and the Economic* Process,[37] that the entropy law applies to entire economic systems, Ostwald had emphasized the general sociological significance of this process. He described it as a duty of culture in all its manifestations 'to design the transformation efficiencies of energy to be as favourable as possible': in other words, to do everything to reduce losses as much as possible through optimal conversion. He ascribed to this Energy Imperative a greater philosophical importance than to Immanuel Kant's Categorical Imperative, which says that one's own maxims should always serve as the basis for the general making of laws. Kant's Imperative targets a moral law, Ostwald argued, which generally causes suffering in human coexistence because it is usually not adhered to. On the other hand, the Energy Imperative is a natural law that 'is being obeyed as soon as it is understood, because there is no possibility outside the law'. If it were not obeyed, human life would one day cease to exist. Based on what is known today about our technical potential, the Energy Imperative must, of course, refer not only to improved energy utilization but more broadly to the issue of how to avoid altogether the use of unexploited energy on the planet. Even when utilization is improved, producing lower conversion losses, this can only delay but not prevent the suffocation of air-breathing organisms caused by emissions.

The Second Law of Thermodynamics is valid for every energy conversion process – although, depending on the energy form, the effects differ widely. The destructive effect is not limited to the mobilization of bound fossil energy in the form of CO_2 or chlorofluorocarbons (CFCs), as is apparent from the destruction of the earth's atmosphere and the ozone layer. It also applies to atomic and nuclear energy; these developed from the findings of high-energy physics that matter is also a form of energy that can be converted into unimaginable amounts of kinetic energy. The splitting or fusion of atoms has dramatically increased the possibility of producing entropy in nature – most drastically with nuclear weapons. Barry Commoner describes civilian use of nuclear energy as 'thermodynamic overkill'[38] because it calls for the generation of temperatures many times higher than are actually needed. The radiant energy from nuclear processes impacts on nature in a way that it cannot cope with; once released, radioactivity diffuses in such a way that it can no longer be contained. Advocates of nuclear energy stress that, in contrast to the combustion of fossil fuels, these processes have no negative effect on the climate. 'Uranium leaves the earth cold' is a slogan used by the German utility industry, for example, in a nuclear energy advertising campaign. But even that is false, proved by the biologically significant warming of the rivers used to cool nuclear reactors.

The 'World Law', as postulated by Christian Shütze, based on the Second Law of Thermodynamics, is valid for closed systems.[39] 'Only under these conditions', writes the German physicist Peter Kafka in *The Basic Law of Ascendancy*, did the law of entropy predict the 'decline':

> The validity of the Second Law assumes not only spatial and material, but also energetic isolation. The figure of Gaia, the earth's surface and her joint biosphere with the sun and the dark night sky as source and sink, surely is not subject to this Law.[40]

In its relationship with the sun, the biosphere is an open system through which the sun's energy flows. The increase in entropy takes place on the sun. The energy radiated from it already provides 'free energy', which means no other energy is needed to harvest solar energy; it is in flux before it is needed. It is, as Barry Commoner says, 'pure energy'. Natural solar radiation warms the land and sea surfaces, where it is absorbed and radiates back in the form of heat. To be sure, the re-radiation is not complete. The sun transforms its energy through plant photosynthesis: food and biomass as indirect solar energy. In addition, the sun does not shine uniformly on the earth: clouds, ice and snow cover reflect and absorb very little sun, daytime and night-time differ according to the seasons and latitude, and an uneven surface changes the angle of incoming radiation. These differences between warmer and cooler radiation zones cause air circulation: wind power as indirect solar energy. Where the sun shines on bodies of water, it causes evaporation of water, which condenses in the atmosphere, falls again to the earth as rain, and flows from higher geographical regions to lower: the flow of water is indirect solar energy.

Of course, the laws of thermodynamics also apply to solar energies, but they do not create additional entropy above and beyond the naturally occurring entropy due to diffusion and reflection into space. On the other hand, there is an increase in negentropy through photosynthetic solar energy storage – that is, plant growth. The flow of energy released by solar radiation thus reduces the entropy by constructing a new order. Solar energy is an integral element of the approximate state of equilibrium of nature on earth. This, too, is not static but subject to constant natural processes of change. By tapping into these solar energies, we integrate ourselves as another user into the existing conversion processes and thus do not endanger the ecosphere, which has been sparked into life by solar energy. We merely delay its diffusion into energy that is no longer useful. Conversion losses accumulated in this process do not release additional heat on a global scale.

When using solar energy, the main difference between good and bad conversion is an economic one: low-efficiency technologies will be expensive and require more material; high-efficiency technologies reduce the capital and material requirements per utilized energy unit. In any case, only a fully solar, global energy economy can preserve the ecosphere. The Energy Imperative, updated by present-day technical knowledge and the potential of solar energy use, should now state: 'Use only those energy sources that do not produce additional heat, emissions and other residues.' In short: 'Use only solar energy sources!'

The Failure of the Century

It is incomprehensible why, over the course of decades, no radical scientific and political conclusions have been drawn from knowledge of the entropy laws. During the 20th century, political initiatives to save energy were launched only when economic crises or wars restricted energy supplies, or when nations had balance of payment difficulties. It is even harder to excuse the decades of ignorance and negligence when the question of how to harness solar energy by using the technological resources of the industrial age was not even asked. In 1926, the French physicist Alphonse Bergeret demanded that solar energy use be the

> task of the physicists and engineers of tomorrow: our coal supplies are gradually drying up, the oil reserves even faster, and it is foreseeable, especially if industry demand continues to grow so

alarmingly, that we will be forced to draw, in a few centuries – perhaps even in 150 or 200 years – the energy we need from the forces of nature. We have water power and tidal power. Wave power is the more important one, and it exists in two forms that we must utilize: the energy of the wind and that of solar radiation.[41]

There were more than enough starting points. The French scientist Antoine Becquerel discovered in 1839 that thin films made from semiconductors produce electric currents and voltages when illuminated, varying in intensity according to the level of illumination.[42]

As early as 1869, the French inventor Augustin Mouchot published the book *Solar Heat and its Industrial Applications*, and introduced the first solar heat-driven steam engine at the 1878 Paris World Fair.[43] But all this was forgotten, just as a solar industrial water heating system that operated in the United States at the end of the 19th century was forgotten.[44]

In 1872, a sea water desalination plant was powered by solar energy in Chile, and in 1913 a 50 hp solar-driven water pump was installed in Egypt that channelled Nile water to the fields.[45] Thousands of well-functioning, small, run-of-river power stations were built in the course of the 19th century. Most have been closed in recent decades because competing large power plants could deliver continuous electricity at lower prices without having to cooperate with small operations.

Hundreds of thousands of windmills were shut down in the 20th century. No systematic attempt was made to develop them further to produce electricity, in spite of the fact that a Dane, Paul La Cour, had already installed in 1891 the first wind power plant for electricity production, in combination with electrolytic hydrogen production, and used it to light a school building. There were 120 working wind power plants in Denmark during World War I, and there was a boom in the United States in the 1920s and 1930s with the construction of about 6 million small plants. Even in Germany, electricity was generated in the 1930s by some 3,600 wind power plants. Admiral Byrd had a wind power plant installed in the Antarctic in 1932, which operated under extreme weather conditions and without any need for servicing until 1955.[46]

There were numerous examples of biogas plants in Germany in the 1940s and 1950s, often constructed by the farmers themselves.[47] The wind power plants went the way of the small hydro plants: as single systems, they were displaced by more efficient power producers that, by installing electric grids, made a more convenient electricity supply possible, while biogas plants were displaced by inexpensive fuels. All these examples of solar energy supply already met the basic criterion, namely that they converted more energy than was needed for their production.

Even France's notable scientific and technical advances in solar thermal energy conversion during the 1950s fell into oblivion. Under the scientific leadership of Marcel Perrot, who published the book *The Golden Coal* in 1962 and who introduced the term 'heliotechnology', revolutionary progress was made at the Institut d'Energie Solaire at the University of Algiers.[48] This in turn led to the Mediterranean Cooperation for Solar Energy (Cooperation Mediterranienne Pour l'Energie Solaire – COMPLES). French advances in this field in the 1950s resulted in a 50 kW concentrating solar furnace capable of generating temperatures of 4,000 °C in Le Bouzereah.[49]

The first solar cell was developed in 1954. Werner von Siemens had alluded to the concept as early as 1875,[50] and its development was visualized by Ostwald in 1909 in these words: 'When I want to create an image of the future artificial utilization of solar radiation, it takes on the characteristics of a photoelectric apparatus'.[51] That first cell was not developed for use on earth, but as an independent electricity generator for space flight.

The first international Solar Energy Congress was held at UNESCO headquarters in Paris in 1973 under the motto 'The Sun in the Service of Mankind'. At the conference, the rocket pioneer Wernher von Braun pointed out that the solar era was the real future.[52]

It became obvious that the technical possibilities demonstrated and discussed at that meeting were suitable for an ecological-technical programme for the future, especially after the Club of Rome produced its spectacular report on *The Limits to Growth* in 1971, followed in 1972 by the first global United Nations conference on the environment, which stirred public opinion worldwide. In 1974, the first solar energy research and development programmes were launched in earnest – though the principal moving force was not the environment but, rather, the oil crisis of 1973, which called into question the permanent security of oil imports. At that time, almost everybody believed nuclear power held the greater promise. But in 1977, the first solar thermal power plant, with a 1 MW output supplying current to the commercial grid, was put into operation in the French Pyrenees near Font-Romeu.[53]

In 1975, the US Federal Energy Agency submitted a plan to start mass-producing solar cells, with the aim of reducing solar electricity prices by a factor of twenty within five years. The plan was predicated on the installation of 152 MW total output at a cost of $440 million, which would replace one fifth of the gas generators used by US military forces, saving an estimated $500 million in fuel costs.[54]

But when the 1978 US Law for Research and Development authorized President Carter to move ahead with the project, he backed off. However, Carter was one of the few heads of state with an open mind on solar energy. On the occasion of International Sun Day in the United States on 3 May 1978, he said: 'Solar energy works. We know it works. The only question is how to cut costs'.[55] The Council on Environmental Quality in the Executive Office of the President that he appointed estimated in a 1978 study that, by the year 2000, solar energy could account for 23% of energy use in the United States.[56]

The political sun first rose, metaphorically speaking, in the second half of the 1970s, boosted by tax incentives and research programmes. The prospect of a 'Solar America' was formulated. Sales by the solar heating industry rose from $25 million in 1975 to $260 million in 1977. A total of 3,300 solar space heating systems had already been delivered, as well as 63,000 solar hot water systems and 35,000 solar-heated swimming pools. At the end of 1978, there were 30,000 solar thermal installations in California alone.[57]

At the end of the 1970s, the entire technical spectrum of solar energy and its economic prospects had been scientifically explored and was ready for business, as documented in the 1979 UNESCO study by Wolfgang Palz, *Solar Electricity: An Economic Approach to Solar Energy*. In 1978, the Groupe de Bellevue – a group of scientists from the Centre National de la Recherche Scientifique (CNRS), the Collège de France, Electricité de France (EdF) and the Institut de la Recherche

Agronomique (INRA) – published a study of France's energy future. The document demonstrated that France's entire energy supply system could be based on solar energy alone.[58]

In 1981, the UN Conference on renewable energies took place in Nairobi and produced the Nairobi Programme of Action.[59] That same year, the then newly elected French President Mitterand started a new solar energy programme and established an agency for it. Although recognized almost negligently late and introduced in fragmented fashion, there were still programmes for research and development of solar energy technologies in most industrialized countries at the beginning of the 1980s.

More recently, this 'sunrise' was followed by twilight: instead of further progress a decade of politically motivated demolition followed. The introduction of solar energy, which had almost been within reach owing to a surge of technological developments, was slowed down by increasingly weaker political efforts. The impression grew that an unwelcome technology was making its debut. Government research budgets were reduced – most drastically in the United States with the advent of the Reagan presidency. The Nairobi programme never got off the ground, and the new French programme was aborted shortly after the establishment of the country's solar agency, and was completely discontinued in 1986.

Those same decision-makers who accuse the critics of large-scale (civilian or military) technological projects of being techno-pessimists act as the ultimate techno-pessimists themselves when it comes to solar matters. Disinformation is spread about the efficiency of solar energy technologies, and positive results are downplayed. So we continue to live in a state of permanent contradiction, suspended uncomfortably between the undeniable scientific evidence of the destruction of the environment and the obstinate refusal of the overwhelming majority in politics and industry to appraise seriously the opportunities of solar energy.

The dawning of solar energy use from 1973 onwards was followed by encroaching darkness in the 1980s. This reveals a most worrying inability of our political and economic representatives to prepare to meet the challenges of the future. As there was not yet any public pressure, owing to the lack of awareness of alternatives, the détente in the oil markets in the first half of the 1980s was taken as justification to drop such new ideas. The old agenda was restored. Business as usual reigned once more. It was not until the Chernobyl reactor disaster of 1986 that this process of suppression was interrupted. It caused many people to give serious thought once more to the idea of a future without nuclear energy, and to revisit the arguments of the anti-nuclear movement, which had been formed in the 1970s. However, the concept of solar energy was so battered that people focused primarily on more economical and efficient use of fossil fuels. Attempts to direct attention towards renewable solar energy sources, which were driven by fears of the climatic dangers of fossil fuels, enjoyed little success. In some quarters of the environmental movement such attempts were criticized on the grounds that they risked legitimizing nuclear energy anew, because it was impossible for solar energy to provide comprehensively for all our energy requirements. The European Solar Energy Union (EUROSOLAR) was founded in 1988 with the express aim of replacing nuclear and fossil energy with solar energy as the 'task of the century'. At the time, energy experts of every hue tended to greet this endeavour with a patronizing smile.

However, then the evidence of climate research about the dangers to the world even from fossil energy became impossible to ignore. Questions about alternatives to the nuclear/fossil energy system grew more frequent. Public curiosity about solar energy became more widespread. Nevertheless, there was little change to the ignorance displayed by the protagonists of the established energy debate. At the 'Our Common Future' conference of the UN in 1990, in Bergen in Norway, solar energy still featured only marginally. Even in the programme debates at Rio in 1992 the government officials writing the text still managed to play down the solar option. When the first edition of this book, *A Solar Manifesto*, appeared in 1993, proposing the theory that renewable energy sources pose a complete and comprehensive alternative, even most solar researchers considered it an exaggerated claim, let alone the opinion leaders of the existing energy system. In the United States, too, the similar theories of Barry Commoner from the 1970s remained ignored.

Despite all this, the possibilities of the solar option for the future unfolded anew in the 1990s. The quest for alternatives became too widespread and too urgent. Independent thinkers were not satisfied with the answer of nuclear and fossil energy options. The solar option forced itself relentlessly into public awareness. Politicians started to include it in their speeches. Politicians and entrepreneurs began to get their minds around this alternative and to develop initiatives. However, the well-known and rehearsed objections are far from being eliminated. Nor has political denial of the future been overcome. In words, but above all in deeds, the priorities of those who denounce the soothsayers as doom-mongers prevail, thereby diverting attention from the fact that it is actually they who are the true horsemen of the Apocalypse.

Renewable energy sources, which were never mentioned for decades in most of the debates about energy provision, have reappeared on the agenda in recent years. Now they feature in almost every discussion. Growth in the number of publications and conferences on the topic has taken off. Similarly the efforts of researchers and developers to find creative applications have increased. Both companies and politicians are showing increasing interest in renewable energy sources. The official policy on renewable energy published by the EU Commission in November 1997 professes faith in their comprehensive ecological and even economic advantages, not least in respect of new job opportunities. Even so, the apologists of the existing energy system, while certainly mentioning renewable energy sources, are in fact increasing their efforts to downplay their true possibilities. In this way they hope to perpetuate the nuclear/fossil energy system and with it their own existence, as well as to justify the previous state of near total ignorance.

In the meantime, political programmes promoting development of renewable energy sources have become commonplace. However, they are still predominantly full of political hype and therefore remain a part of the failure to prepare for the future.

The Failure of Politics to Meet Future Challenges

A lack of tangible opportunities and, in particular, lack of money, are the usual reasons for the refusal to accept the prospects for the use of solar energy. A comparison of the numbers illustrates graphically how core tasks for the future are ignored in grotesque fashion in favour of other priorities. It has frequently been pointed out that political reluctance to get involved with solar energy has been disproportionately high compared with the decades-long massive political push on behalf of nuclear energy. This is by no means the only instance of alarming political priorities in development of technology, as demonstrated by the relevant figures from the budgets of various governments. In addition, there are the numerous financial incentives, above all in the form of subsidies and tax relief for non-renewable energies, which far exceed all previous political initiatives.

Example: Energy Research

Table 3 on page 49 shows the total expenditure from 1984 to 1995 for research and development of renewable energies by countries in the Organization for Economic Cooperation and Development (OECD) or the International Energy Agency (IEA). The overall total expenditure amounted to $9.277 billion at 1995 price levels.

The annual total in 1984 was $1.098 billion. By 1989, this figure had dropped to $619 million. The decline was similar in virtually all countries. Taken together, the IEA member nations' budgets for 1989 were only 56% of the 1981 level. Energy research activities in general declined after the second oil crisis at the beginning of the 1980s. Nevertheless, it is obvious that support for solar energy in particular was kept deliberately small, even though it is the newest and most wide-reaching area of publicly supported energy research, even more dependent than other areas on public funds and starting from the lowest level. Even energy conservation and efficiency, the darling of environmental energy-speak, was trimmed back more than the fossil energies. What was already a meagre $753 million in spending in 1984 fell even further to a pitiful $482 million in 1989.

The imbalance of these sums becomes apparent from the International Energy Agency statistics in Table 4 when we compare the total expenditures for energy research in the IEA countries between 1984 and 1995. In 1984, the research and development expenditure on renewable energies of the IEA member countries amounted to $1.1 billion, obviously still under the influence of the oil crisis. Thereafter, it declined

Table 3 *Research and development budgets of IEA countries for renewable energy in millions of $ (1995)*

	Year											
	1984	1985	1986	1987	1988	1989	1990	1991	1992	1993	1994	1995
Canada	53.3	31.6	18.4	15.7	14.3	11.8	9.8	9.4	10.8	9.7	11.3	11.0
USA	317.1	289.7	213.9	194.2	152.0	133.5	123.1	163.5	228.1	219.0	329.4	393.0
Japan	219.4	197.1	192.2	170.9	188.7	154.8	153.5	150.1	144.0	194.4	138.0	139.4
Australia	–	10.7	–	1.0	–	4.7	–	–	–	–	–	–
New Zealand	5.7	4.1	1.8	0.6	–	–	0.5	0.5	–	0.9	1.0	1.1
Austria	6.4	6.2	4.6	4.8	7.3	4.0	2.7	6.1	5.2	7.1	7.6	–
Belgium	21.7	20.9	8.2	6.3	3.0	0.7	–	–	–	1.7	9.2	–
Denmark	4.9	4.9	6.6	5.9	–	12.9	10.4	21.1	22.5	24.1	20.9	18.6
Germany	135.3	120.8	75.5	106.2	113.0	110.8	131.7	144.1	151.3	165.1	109.2	122.1
Finland	–	–	–	–	–	–	10.4	2.3	13.2	12.5	–	–
France	–	–	–	–	–	–	17.0	9.7	9.4	6.7	6.3	6.1
Greece	7.0	10.6	16.8	8.1	20.1	9.4	4.6	4.7	5.1	3.7	–	–
Ireland	1.4	1.0	0.8	1.3	1.5	–	0.5	–	–	–	–	–
Italy	101.4	26.6	43.1	39.1	54.9	42.3	48.4	36.9	–	27.3	30.8	42.0
Netherlands	28.5	67.1	32.6	29.5	24.4	28.3	42.7	41.6	24.7	24.6	21.4	20.2
Norway	4.2	3.7	3.9	2.8	2.8	3.5	5.9	10.5	11.6	9.2	7.5	4.9
Portugal	5.0	4.9	4.1	3.2	2.8	4.0	2.1	2.0	2.6	1.7	0.6	0.6
Spain	75.4	24.9	21.5	14.3	15.5	16.4	21.9	18.3	25.0	22.5	16.5	65.6
Sweden	57.5	39.3	27.1	20.1	22.5	22.7	19.1	12.7	30.9	15.5	18.6	14.4
Switzerland	19.6	16.9	16.7	18.6	24.7	30.4	32.5	33.3	53.4	49.0	46.8	43.1
Turkey	0.9	0.7	0.9	0.8	1.2	0.9	0.2	0.1	1.2	0.3	0.4	0.4
England	33.4	28.9	23.2	27.7	28.4	28.3	27.8	30.3	27.3	24.9	14.6	14.3
Total	1098.2	910.6	711.8	671.0	676.8	619.3	664.8	697.3	766.3	774.7	790.2	896.8

Source: 'Energy Policies of IEA Countries', 1995, Review, Paris: OECD

Table 4 *Annual research and development budgets of IEA member countries by energy source (in $ million at 1995 prices)*

Year	Renewable energies	Oil and gas	Coal	Conventional nuclear energy	Fast breeders	Nuclear fusion	Energy conservation
1984	1,098.2	536.4	1,179.5	4,395	2,166.5	1,523.8	753.2
1985	910.6	434.2	1,180.1	4,662.3	2,045.1	1,509.8	740.4
1986	711.8	548.8	1,063	4,333.5	1,772.6	1,390.2	632.6
1987	671.0	492.3	969.8	3,512.2	1,128	1,315.3	686.8
1988	676.8	389.7	1,154.4	2,592.6	1,276	1,208.1	571.2
1989	619.3	341.7	1,032.1	3,353.1	1,152.8	1,138.2	481.8
1990	664.8	386.8	1,442.1	3,786.6	1,023.1	1,163	588.9
1991	697.3	412.1	1,179.1	3,878.9	898.8	1,064.4	630.4
1992	766.3	404.7	732.1	3,392.7	913.3	1,035.5	592.2
1993	774.7	460.2	676.6	3,383.6	731.4	1,121.6	694.2
1994	790.2	496.2	837	3,428.9	624.7	1,073.5	966.8
1995	896.8	560.8	578.4	3,675.9	473.8	1,105.1	1,023.5

Table 5 *Research and development budgets for energy sources from 1984 to 1995 (in $ million at 1995 prices)*

	Energy Conservation	Oil and gas	Coal	Conventional nuclear energy	Fast breeders	Nuclear fusion	Renewable energies
USA	3,201.3	1,206.9	5,142.7	6,893.5	466.3	4,979.1	2,756.8
Japan	1,048.6	1,842.2	4,281.8	25,729.2	8,116.5	4,849.6	1,997.5
Germany	280	176.4	1,391.5	3,110.5	1,485.5	1,930.8	1,485.1
Austria	139.3	11.4	6.7	27	0.5	25	61.7
Switzerland	318.1	93	7.8	415.2	45.5	335.7	385
Great Britain	458.6	284.2	109.9	860.5	1,607.1	499.7	309.1
France*	150	253.9	39.3	2,778.7	360.8	288.6	55.2
Sweden	420	14.9	95.1	16.5	51.9	132.6	300.4
Netherlands	690.3	57.8	263.7	337.1	15.2	208.9	385.5
Italy	568.8	0.8	35	2,158.6	1,695.7	1,048.3	492.8
All IEA Countries**	8,362	5,472.9	12,015.2	42,326.8	14,106.1	14,647.9	9277.8

* Figures from 1990 to 1995 only

** Canada, USA, Japan, Australia, New Zealand, Austria, Belgium, Luxembourg, Denmark, Finland, France, Germany, Greece, Ireland, Italy, Netherlands, Norway, Portugal, Spain, Sweden, Switzerland, Turkey, Great Britain steadily, instead of rising further: by 1989 it had fallen to 56.4% of the 1984 total. It then gradually picked up. It was only in 1995 that it got back to 81.7% of the 1984 level. Even in the area of energy efficiency, to which everyone pays lip service and which is therefore indisputably essential, financial support fell between 1984 and 1989 from $753 million to $482 million. It was not until 1994 that there was a move upwards. In 1995 the amount of $1.023 billion reached 136% of the 1984 level. It is interesting to note by contrast that spending on nuclear energy research has by no means suffered the enormous reductions that are so often claimed. For three decades from the mid-1950s onwards, nuclear energy was the firm favourite of government spending policy. In 1984, $4.39 billion was still being spent on research and development of conventional nuclear energy, $2.16 billion on fast breeders and $1.53 billion on nuclear fusion. In total that was $8 billion, just about eight times more than spending on renewable energies. In 1995, it was still $5.25 billion, almost six times more than for renewable energies. This apparent cut was due predominantly to the massive reduction in support for fast breeders, whose unsustainability could no longer be concealed.

Table 5 on page 51 gives the breakdown of publicly funded R&D for each energy source by country. Note that nuclear and fossil energy also benefit from privately financed research by large business sectors, resulting in an even greater mismatch of future-directed efforts – which is where research should be aimed – at the expense of solar energy. In addition, French and European Union budgets are included in these IEA energy research statistics only from 1990, further raising the nuclear research share. Furthermore, the EU commission support quotas are not included in the IEA statistics. This, too, tips the balance further in favour of nuclear energy. It is a myth that the absolute precedence given to nuclear research was just a phenomenon of the 1950s, 1960s and 1970s. It remains a mystery why governments still support fossil energy research at all, in view of the gigantic turnover of corporations in these areas. Instead of equalizing this mismatch with solar energy through government-funded research – which should be the job of government policy – solar energy has been neglected further.

As this has been happening in virtually all countries, it leads one to conclude that it is due to a colluding international opinion Mafia.

Although no more nuclear power plants have been built in the United States since 1973, $6.9 billion was spent on research and development for conventional nuclear energy between 1984 and 1995, as opposed to $2.7 billion for renewable energies. Even coal received nearly double the amount for renewable energies. If we add up the research and development expenditure for conventional nuclear, fast breeders, and nuclear fusion, we reach a total of $12.33 billion versus $2.75 billion for renewable energies – that is, a ratio of 4.5:1.

The imbalance to the detriment of renewable energies is at its most extreme in France. Between 1990 and 1995 expenditure on nuclear energy totalled $3.42 billion against $55.2 million for renewable energies – a ratio of 62:1!

In Japan, too, nuclear research is held in higher esteem than renewable energies, which is contrary to expectations, but not entirely surprising given their highly centralized research and development support programme. True, Japan spent about $2 billion on renewable energies between 1984 and 1995, but $38.7 billion on nuclear research, a ratio of 19:1 in favour of the latter!

It is remarkable that Italy continued with nuclear energy in spite of the fact that there was a successful referendum against it in 1987. From 1985 to 1990, it spent $4.9 billion on nuclear and $492.8 million on renewable energies: a ratio of 10:1 in favour of nuclear energy! Great Britain has also hardly changed its priorities, in spite of its difficulty in selling off nuclear power plants during the privatization of electrical utilities. It spent $2.96 billion on nuclear energy compared with $309 million on renewable energies – that is, again almost 10:1! In Sweden the decision against nuclear energy has been taken seriously. It is the only country that once set store by nuclear and yet has invested more in research and development into renewable energies between 1984 and 1995.

Taking all IEA member countries together, the expenditure on research and development policies from 1984 to 1995 is in the following order (in $ billion):

Conventional nuclear energy	42,326
Fossil energy	17,487
Nuclear fusion	14,647
Fast breeder reactors	14,106
Renewable energies	9,277
Energy conservation	8,362

The numbers for nuclear would be even higher were it not for the fact that during this period, especially in the United States, Germany and also in France, fast breeder research collapsed. They would also be higher in fossil energy but for the fact that research and development for oil and gas have long been predominantly pursued independently by private enterprise.

Research and development expenditure does not tell the whole story. However, it is an indicator of governments' policy. What it shows is that, time and again, taxpayers' money is ploughed into development that runs counter to the best interests of the taxpayers' own future.

The imbalance in favour of false priorities – and with it the potential room to manoeuvre for new priorities – is more obvious from a comparison of other government-supported technologies with the renewable energy research efforts.

Example: Space Research

More than $500 billion was spent worldwide on space research up to the end of the 1980s, approximately 50 times as much as on solar technology. The centres were, initially, the Soviet Union and the United States, but for two decades Europe has been playing catch-up. There are also Japanese and Chinese programmes, as well as Canadian and Indian ones.

Ever since the Soviet *Sputnik*, the 1950s' technological Wonder of the World, space technology has been a symbol of global influence and technological progress. The United States focused its main political ambitions in the 1960s on efforts to close the Soviet lead in space technology, something that was regarded as a national disgrace. President Kennedy's Apollo programme became a national mission. The European states regarded themselves as inferior, with no programmes of their own. By exploiting, with the tools of mass psychology, the public's fascination with humans landing on the moon or circling the planet for months, space research became the number one priority of civil research policies; objectors were labelled as hopelessly provincial and behind the times. Europe's governments founded the European Space Agency (ESA), consisting of thirteen member states (Austria, Belgium, Denmark, France, Germany, Great Britain, Ireland, Italy, the Netherlands, Norway, Spain, Sweden, and Switzerland). Its long-term plans incorporated all the projects that the United States and the Soviet Union had in their programmes – from satellites to manned and unmanned space craft – regarded as future basic equipment of a modern industrial society.

For the period 1987–2000, ESA planned a $38.7 billion programme (at 1986 prices) – about $2.75 billion annually.[60]

Today there are complaints that these funds are not sufficient, and calls for greater spending are becoming more urgent. A comparison of the contributions of these thirteen countries alone for the long-term space programme with those for solar energy in 1995 shows a ratio of 5:1 in favour of space technology. This is only considering funding for ESA; the true ratio is even more favourable for space technology when national efforts are included. Including national budget financing, $4.38 billion is being spent year after year by ESA's member states for space research – eight times as much as for solar research and development.

However, mankind's real problems are here on earth and not in space – and it is certain that the jobs of the future will be found in solar technology, not in space technologies. So far, however, the ESA member states are spending more on the technological observation of environmental damage than on solar technology to prevent environmental damage.

Example: Weapons Technology

The comparison of priorities becomes even more dramatic if we look at the figures for large military projects, based on current military procurement programmes, no longer justified by the military threat from the Soviet Union. These programmes are carried out in an international situation defined by the military technology of the United States and the key NATO countries, vastly superior to that of any conceivable military opponent. Here, too, it is always a matter of 'national missions' that overshadow any question of how they compare with others, or of their ecological-industrial purpose.

Let us just call to mind the international discussion about the USA's Strategic Defence Initiative (SDI) programme for missile defence in space, which was started in 1983. More than 50 of America's science Nobel Prize winners declared that SDI was technologically impossible and economically irresponsible, but the response to these critics was that they were hostile to technology as such. The governments of Western Europe parroted that line of reasoning, and attempted in near-panic to become part of the development team, arguing that European technology would otherwise deteriorate to the provincial level. These pronouncements came from the same politicians who had always maintained that the utilization of solar energy was 'unrealistic'. The same practitioners of *realpolitik* swallowed the idea of building, in space, a nuclear power plant and radiation mirrors to provide power for laser beams aimed at incoming missiles via mirror systems thousands of miles apart! In 1986 I asked why no government declared an ecological SDI programme – a Solar Development Initiative – as a national mission that would inevitably be technologically less demanding and more realistic in its execution.[61] For the same reasons, then Senator and recent Presidential candidate Al Gore called for a 'Strategic Environment Initiative' (SEI) in 1990.

SDI has since been modified, and its technical demands brought more down to earth, but it is nevertheless being continued, even after the dissolution of the Soviet Union. It is an open secret that the issue is no longer one of security and new military technology, but rather a large-scale procurement programme for the survival of the defence industry, whose technology, it is claimed, will have a large civilian spin-off. A total of $29 billion was spent between 1983 and 1991 for the SDI programme. In the same period, the US Federal government could afford only $1.4 billion for its solar technology programme – a ratio of 20:1 in favour of the most expensive armaments programme of all times, compared with the most important future energy technology in the world's history. The US government spent over $40 billion between 1992 and 2000 to continue the SDI programme. It is not only this plan that raises fundamental questions, nor is it solely the American government that behaves with manic obsession, with its extreme preference for militarily senseless projects, while it neglects urgently needed civil technology projects.

The bottom line: at $393 million, the United States government in 1995 spent less on solar technology development than the cost of one SSN 688 Los Angeles class warship for submarine combat (at about $500 million each), and only a little more than one B1B bomber (at about $300 million apiece).

Let us imagine that the United States commissions only half the 62 warships and half the one hundred B1B bombers it originally intended, and puts the money saved into development of solar technology. This would represent an amount of $30 billion. Not even the most paranoid American security fanatic could argue that this would endanger military security, especially these days. However, because of political ideology and collusion of interests, the American leadership has failed to recognize so far that its claim to world leadership means the nation must embark on an all-pervasive conversion of the armaments business and a technological orientation towards environmental safety.

Britain's programmes have similar absurdities. For example, expenditure on the Trident programme for new submarine nuclear platforms and nuclear-tipped missiles totalled nearly £15 billion. This alone is around 15 times more than the UK government ever spent on the development and introduction of renewable energies. Following announcements by the new US President, George W. Bush, the NATO states are currently discussing their participation in the Nuclear Missile Defense programme, the revised version of the SDI programme. Yet again there is a danger of setting wrong priorities.

Economic Fetishes and the Pessimism of Expediency

Other comparisons, too numerous to mention, such as that between research into biomass energy and spending on agricultural subsidies for the tobacco industry or for food exports from the European Union (EU), could be added. However, the rapid sketch provided by the examples above is sufficient to confirm how little governments react to the environmental and economic demands of the future. It shows how underdeveloped is the willingness to take a self-critical look at priorities once they have been set. It also demonstrates a few other things: that there is such a thing as international pack behaviour, despite questionable pack leaders; that research essentially serves only to prolong priorities of the past rather than to cope with the future; and that there is an inability to take a real step to correct erroneous developments once they have been recognized as such – in short, an inability to reshape the community.

The explanations for sticking to the established, politically motivated, technological priorities betray a threadbare pessimism born of expediency, and do not stand up to plausible counter-arguments. For example, the argument for lopping off support for solar technology was based on the new decline of oil prices that began in 1983 – as if oil price fluctuations ever had any influence on public R&D programmes in nuclear energy! Another argument was that there was no consensus in society supporting solar energy – but most military projects are pushed through despite social dissent, and governments attempt, with huge public relations efforts, to produce public consent for weapons. Nevertheless, there have been, and there continue to be, mass protests against military projects and nuclear power plants, while a mass demonstration against a solar power project is difficult to imagine. What is meant evidently is the lack of consensus among the elite, whose points of view count for more than that of the general public even though governments are supposed to abide by democratic principles.

The often-repeated argument that there is a lack of attractive solar energy projects that would legitimize public support is especially skimpy. For one thing, such a deficiency should not prevent any government from launching a market introduction initiative in lieu of R&D support, something that has seldom been attempted. For another, that argument is simply an impudent excuse for political passivity: once a government recognizes the general political significance of solar energy, it should not have to wait for suggestions from the few scientific and technological institutes: it must define development goals on its own. The spectrum of research and development tasks was and remains so large that, independent of the need for immediate market introduction, a broad and large-scale task presents itself to science and technology – from basic to applied research. A glance at the reports from international photovoltaics, solar thermal, wind energy and biomass conferences provides enough stimulation to formulate an extensive research programme.

Finally, it is always objected that there must be more research and development first: that the time is simply not yet right for market introduction. This statement directly contradicts the previous one, that there are not yet sufficiently promising research ideas, and is in any case plainly wrong. The spectre has been raised that it may be a mistake to start mass production of a technology that may later prove to be outdated. The production of TV sets was not deferred until colour TV was available, nor did the production of radios wait until the advent of multi-frequency receivers! None of these products would have made it to the market, if arguments of this type predominated before their launch.

In reality, it is the feigned naivety of industry that lurks behind these arguments. Today, the launch of a new car model entails the outlay of several billion dollars for its development. As with every other technology, the development of solar technology will accelerate only if and when significant production and market experience has been accumulated. This is illustrated by the example of the Danish and German wind power industry, which was given a considerable boost in terms of quality in the 1990s merely from the newly introduced electricity supply regulations in those countries. As soon as the market opened, technology and industry developed much more quickly than most people had anticipated. Governments should have a strong self-interest in starting industrial volume production as early as possible, because the sales thus generated would provide the financial elbow room for further development and improvement of the technology and further production increases in the private sector, and thereby the possibility of those much coveted and much acclaimed new job opportunities.

The fact that solar energy is treated with such small-mindedness is a sign of the backwardness of Western decision-making elites. Instead of formulating a political future appropriate to scale and problem, government shows of razzmatazz are staged, because in the glitzy post-modern media age, palace reportage sells better than genuine reporting of genuine problems. Officially, all governments have now recognized the fateful implication of the change in the planet's atmosphere. Admittedly there are remarkable differences in the way various governments are coming to grips with the problem, but a really adequate strategic action programme, given top – or at least high-ranking – financial priority does not yet exist anywhere, except in Sweden, Denmark and perhaps the Netherlands.

Pathfinder to Mars instead of Pathfinder for Solar Energy: the Faustian Absurdity of Our Age

The accusation that decision-makers are guilty of such a basic omission appears to lack credibility. It seems unimaginable that nearly all leadership elites are capable of making such basic errors of judgement – although they should know better. Nevertheless, historical examples abound of what Barbara Tuchman described as the 'March of Folly'. This folly refers to basic errors made over long periods of time, whose consequences were clearly predictable and for which there existed in each case an actionable alternative. Tuchman wrote:

> A phenomenon noticeable throughout history regardless of place or period is the pursuit by governments of policies contrary to their own interests. Mankind, it seems, makes a poorer performance of government than of almost any other human activity.

Tuchman differentiates among four kinds of misgovernment that may appear simultaneously: tyranny, excess of power, incompetence or decadence, and folly or mental standstill. One may pick and choose which element predominated in the dissolution of the Soviet Union, rendering it incapable of correcting the system in time for its continued survival, and which elements predominate in the West, in addition to excess of power. Tuchman concludes:

> We can only muddle on as we have done in those same three or four thousand years, through patches of brilliance and decline, great endeavours and shadow.[62]

But there is a prime difference between the present and the

past: as mankind's very survival is at stake, we cannot afford any more folly.

In recent years it has been increasingly recognized that technologies must be 'error-friendly' – in other words, we should avoid technologies that do not permit any errors because their harmful consequences would outweigh the means available to compensate for any damage.[63]

This is valid not only for individual technologies, but also for entire social systems. Their 'error-friendliness' is lost once these systems get involved in uncorrectable dangers, ending in a hopeless debacle. Today, we are faced not with just one but with several dangers of this type – all proof of the collectively faulty behaviour of the leadership elites. Nuclear armament is spreading further and further across the globe the longer nuclear disarmament is delayed. The wider use of nuclear power for supplying energy spreads, the greater are the dangers of reactor accidents, and the more nuclear wastes will turn into a mortgage that will be almost impossible to pay off in the future. There are claims that the economic consequences of the Chernobyl reactor accident were one of the lethal blows for the former Soviet Union. The use of fossil energies – as we are increasingly aware – is part of that: some insurance industry studies regard climate changes caused by fossil fuels – such as severe storm damage – as no longer insurable in the future.[64]

Also included in this category are the uncontrollable misuse of genetic engineering, and the type of ambitions linked to the 'Star Wars' programme and to space flight. These are Janus-headed concepts, which hold out fascinating, allegedly indispensable, advantages, but at the same time present extremely sensitive and pervasive disadvantages. The first most important characteristic of such technologies is this: their introduction changes society in a totally unexpected fashion, although initially they are perceived as just one more technology. The real or imagined advantages they present to society are disproportionate to the singular, long-term dangers inherent to them. The second characteristic is that these are technologies that make it possible to play with mankind's fate. Since the development of the nuclear bomb, scientists and politicians have evidently learned to think and act impassively in terms of the life and death of all humanity. Fascinated by the fact that they must avoid the unthinkable, they nevertheless cling to the ability to bring it about. Entranced by the breathtaking scientific insights and technological developments provided by nuclear, weapons and space researchers, governments have erected pedestals for them and gratified every one of their wishes in order not to appear retarded and limited. Those elevated to the Olympus of the natural sciences are automatically accorded 'intellectual and spiritual leadership' in society.

Thus the ideology that mankind can or should shape nature according to its needs instead of taking its cue from nature's rules has taken hold. 'A Farewell to Nature' is a slogan that long remained unspoken, until one single author was bold enough to proclaim it. It can be read in a frivolous pamphlet by Ben-Alexander Behnke,[65] in which Nature not only can no longer be preserved, but should no longer be preserved. In an interview on the occasion of the twentieth anniversary of the moon landing, the futurologist Jesco von Puttkamer articulated with rare candour:

> The protectors of the environment – and with our space programme, with 'Mission Earth', we are part of that – are actually still talking about a kind of 19th century ecology if they ignore space. But earth doesn't exist all by itself. Part of it is space now, with its human beings who function and build in space. New interactions must be created by this coming together of space and earth that will open for men the closed system of earth – something that inhibits our future growth – and that provides an alternative to a closed future with its scarcities, with its environmental damage, with the possibilities of wars and nuclear mushrooms. Space travel is the hope that we must have to provide alternative life prospects for the mankind of the future. Some day we will build artificial biospheres in space itself. We have to consider that we can't cope any more with our own biosphere here on earth, that the natural environment does not seem suitable for a race of beings that grows as dynamically as man does – with industry, with

wastes, with its energy consumption. Somehow, the earth's natural biosphere and man are not compatible. Yes, it even seems that we are living in a hostile environment, because otherwise we wouldn't find ourselves in such conflict with it. But it is entirely possible that some day we will be able to build artificial biospheres in space from scratch, closed cycles that are optimized for man – in other words, constructed in such a way that they are ideal for man and no longer in conflict with a dynamically growing entity named humanity.[66]

These sentences reveal a modern master race philosophy that is more interested in the question of how one can escape from a destroyed earth with a new Noah's Ark, than in finding a new accommodation with the life around us on earth. The elite able to reserve the few seats on these spaceships would be the descendants of those who today would risk the ruin of the earth's civilization rather than do everything possible for its salvation. As abstruse as the 'philosophy' of the 'futurologist' cited may sound, the fact is that considerably more money has been spent on technologies for the flight from earth than on technologies for its rehabilitation.

The scientific, economic and political elites continue to cling to the blatant contradiction between the prioritized development of incredible technical capabilities and the simultaneous neglect of incredible threats. Today, they may be at the centre of public attention and honour but in the histories of the future they will appear as gamblers, as mankind's felons – because they are operating with full mental accountability, and they possess the information about the acute dangers faced by the present generation of mankind.

Goethe's *Faust* described the dilemma long ago in clairvoyant, literary form: the attempt to wrest secrets from nature that appear indecipherable and the wrestling for limitless exploitation of the world, in exchange for the willingness to accept limitless problems in its wake. Thus human capabilities have been so overtaxed and man's physical and spiritual limits exceeded to the point that all of mankind is now threatened by agony. 'Because of self-acceleration, the evolutionary process violates its irreplaceable preconditions which I like to describe with the slogans "diversity" and "leisureliness" ', writes Peter Kafka.[67]

It is no accident that the use of solar energy does not fascinate and preoccupy politicians and scientists. There are many Nobel Prize winners in nuclear physics, but the development of photovoltaics, for example, has not been worth a single Nobel Prize or any other major scientific prize to date. This is because the use of solar energy does not generate capabilities that exceed the limits of human nature, but 'merely' enhances and supports the abilities required for a life within these limits. Nuclear, fusion and space technology fit into the philosophy of this hypertrophic age, even when they quite evidently do more damage than good – but not solar technology. Solar energy use is not the only, but probably the most important, technology to lead humanity back to a path within its limits. As long as science and politics do not regard this as the more fascinating task, they are on a path to global destruction.

Puttkamer's science fiction is well on the way to shaping modern patterns of behaviour. This was illustrated by the ideas broadcast in the mass media at the time of the *Pathfinder* mission to Mars in July 1997. It was proclaimed as a 'historic' event for mankind, providing the prospect of transforming a planet 190 million kilometres away into a giant biological environment capable of supporting human life, in order then to be able to colonize it. This would give us the possibility of emigrating to Mars after the destruction of the conditions to support life on earth. It is shocking that hardly anyone – whether in politics, science, the media, or even among ecologists – worked out the topsy-turvy absurdity of such an undertaking. Obviously all too many people have been gripped with anxiety that they will be cast as technophobes and discredited. Everyone clearly wanted to avoid wrecking the illusion conjured up by such a space event. After all, this might risk drawing attention to the fact that however many flights to Mars we make, the future of mankind is right here on earth, or not at all.

Mankind will only really understand the problems of existence properly when it starts focusing its technological missions on rescuing the ecosphere of this planet, rather than on generating conditions for life on Mars. As long as scientists, media and politicians are more interested and involved in the latter, we are dealing with a literally breathtaking collective delusion. The mere idea is in the realms of fantasy. There would be giant solar mirrors in space to warm up Mars, causing water to spring forth, so that, with the CO_2 in the Mars atmosphere, plants could grow and help to produce oxygen. It is just too far-fetched to believe that by generating a biosphere, life for mankind on Mars would be possible. This idea verges on the surreal when it is deemed more desirable or even more realistic than substitution of nuclear and fossil energy with solar energy, with a view to preserving the earth's atmosphere and securing Man's future on this planet.

Whatever you think of space travel, the following cannot seriously be disputed. Mobilizing solar technology for the energy requirements of all earthlings, cleansing our air, water and land, permanent greening and recultivation of desertified and eroded areas must be achievable in a more cost-effective and much less complicated way than a potential reconstruction of conditions for life on Mars. However many mineral ore reserves there may be on Mars, their transport to our planet would cost considerably more than achieving efficient recycling of raw materials on earth.

Even the most popular alibi for projecting human life beyond earth does not stick: the apparently inexorable rise in world population. In societies that have reached relative social and ecological stability, population has long ceased to increase. It is only growing in places where people cannot afford basic necessities – food, housing, education and health provision – because the economy and society are organized in a lop-sided and inadequate way. Who could possibly pretend that this is going to be better on Mars?

Why should we listen to those who are incapable of preserving the ecological conditions that are necessary for life on earth? Let us keep our feet on the ground – earth's ground. Let us speak out vigorously against the ideology of space. The sun favours us, not Mars. The great experiment in sustaining life is here on earth, not on Mars. The desire to colonize Mars using the opportunities provided by ecological science and modern technology is merely an attempt to justify further environmental pillaging. Mankind's missions should be *Apollo* and *Pathfinder* programmes for earth. To keep the green planet green in a natural way, rather than to try to make the red planet green in an artificial way, we need not the *Pathfinder* to Mars, but the *Pathfinder* of the sun for our earth.

3

Are there Alternatives
to a Global Solar Strategy?

Those who call for the use of solar energy are still faced with the basic question of whether this is really a serious alternative capable of providing something more than merely a minuscule share of the total energy supply. However, given the scorched and irradiated earth left by fossil and nuclear energy carriers, to be relevant the question should be reversed: is there any alternative to solar energy?

The view that solar energy should be *the* strategic issue in today's efforts to tackle energy policy and energy economy is held by only a very few individuals so far. The vast majority of those trying to formulate answers to the global climate threat offer other priorities. Two schools of thought are evident: one, the Business-as-Usual School, doesn't want to give up any of the current energy resources. Their idea is to reduce the use of fossil energies by energy-saving strategies, to phase in solar energy gradually, to keep nuclear energy going, and to bet on fusion energy for the long term. The majority in the other, the Energy Conservation School, is willing to abandon nuclear energy, and supports the more rapid introduction of renewable energy. But their priority centres on a strategy of energy conservation, the 'efficiency revolution', and not a solar strategy. Energy conservation is called the largest new source of energy.

There is a general consensus between both schools of thought in favour of energy conservation strategies. This consensus makes the demarcation lines between the schools somewhat fluid. The disagreements are over nuclear technology; the Business-as-Usual School is less dedicated to energy conservation strategies, and wants to hold on to the centralist structures of the energy business. One could object that such a definition of these schools of thought is rather artificial – and even more artificial is the differentiation between these schools on one side and the solar strategy championed here. Everything, it could be argued, is simply a matter of different estimates of the time needed to achieve successes in energy conservation and to launch solar energy in the energy markets.

Such objections fail to recognize that a great deal more is at stake than differences in judging practical possibilities. The energy system is not a value-free and interest-free laboratory. Different interests and goals will lead to different actions. The wrong goals will lead to bad results, insufficient goals to insufficient results, and incomplete goals to the neglect of necessary points of departure. In order to debate the question whether there are in fact alternatives to the strategic goal of a global, completely solar energy supply system, the importance of energy conservation strategies must be examined, leading to the question of the future of nuclear energy, which ultimately leads to the question of fusion energy.

The Limits of Energy Conservation

In all the programmes from political institutions to tackle the threat of a climate catastrophe, the demand to save energy predominates over all other suggestions, either via direct limits on consumption or via incentives to improve energy productivity in the various links of the conversion chain.

Among industrial nations, Germany has the most ambitious official goal, with a 25% reduction of CO_2 emissions by the year 2005. The Inquiry Commission of the Bundestag (German Parliament) went even further, suggesting reductions of 30% for the industrially most developed countries, based on 1987 levels, by 2005, and a 50% limit in emission growth for the developing countries, which on a global scale would achieve a 5% reduction of CO_2 emissions. For the year 2050, a goal of reducing CO_2 by 80% was recommended for industrial countries, and developing countries were to be allowed 70% emission growth, which would lead to a total worldwide reduction of 50%. For the period up to the year 2005, the Commission, which was regarded internationally as a model for studying climatic dangers seriously, was relying primarily on energy-saving technologies: in existing buildings, in new construction, for small users, electrical appliances, cars and airplanes, buses and trucks, hot water production, industrial fuels, cogeneration and in industrial electricity supply. The utilization of renewables would only achieve 'in the period after 2005 significantly larger shares of the energy supply and in the reduction goals'. After that – between 2005 and 2050 – CO_2 reduction via the use of renewables could achieve 'roughly about the same order of magnitude as the one obtained with more rational use of energy'.[68] The argument in the Commission centred not so much on the question of a greater emphasis on solar energy but on whether nuclear energy should be included or not in a strategy to prevent catastrophic climate changes. Here we can clearly see the existence of the two different schools of thought mentioned above.

The World Resources Institute in Washington D.C. optimized the known assumptions about an economically and technically proven reduction of energy demand in a world scenario, taking into account population growth and the minimal requirements of developing countries. According to this scenario, primary energy consumption would increase from 10.3 to 11.2 TW per year between 1980 and 2020 as the world's population grew from 4.4 to 6.9 billion. However, this calculation assumed that the industrial countries would reduce their consumption from 6.3 to 3.2 TW (despite a population growth from 1.1 to 1.2 billion) – roughly a halving of per capita energy consumption – while developing countries would double their consumption to 8 TW by the year 2000 (including a population increase from 3.3 to 5.7 billion).[69]

These figures underline the urgency of the need for a solar strategy, because they calculate the achievable effects of energy conservation strategies optimistically. If further increases in the next thirty years cannot be prevented, even assuming specific 50% energy savings in industrial countries, then much more drastic countermeasures will have to be launched, as by then the climate will already have begun to show signs of anomaly. A provisional strategic limitation via energy conservation would not even get halfway to its goal.

The fact that the Bundestag's Inquiry Commission consciously focused its recommendations on energy savings strategies and not, at least to the same extent, on solar energy is indicated by its remark that it would be possible prior to 2005

to produce in volume small and medium-sized wind converters, hydro-power plants, biogas plants for the fermentation of agricultural waste, landfill-derived methane and sewer gas, and solar low-temperature plants and therefore achieve 'in the short term, limited, but not negligible market shares'. However, it stops short of recommending immediate, massive political initiatives in support of these alternatives. If the Commission sees opportunities for mass-producing these solar technologies, why is it reluctant to take decisive action and to demand immediate, significant efforts to kick-start this process?

Another observation is equally indicative of this unfounded reticence. Plants for the production and utilization of photovoltaics and solar hydrogen systems ostensibly require 'a start-up time of several decades before they can make a significant contribution to the supply of energy', the Commission said. Why do we still need to think in terms of decades, if photovoltaic cells are already being produced today by several dozen companies? Nothing stands in the way of a targeted expansion of production other than the lack of political and economic perspective and determination. To be sure, it is clear that the start-up times are not the same for the various solar technologies because of their respective differences in cost and level of commercial readiness.

Evidently, the Commission allowed itself to be reined in by the widespread objections that an immediate, offensive strategy to introduce solar energy would be too expensive. Such objections are factually wrong, as we shall see in the next chapter. Even if they were economically proven, these arguments could not justify the fact that a strategy urgently needed to save the planet's atmosphere is not being recommended simply for reasons of cost. The German Parliamentary Commission proceeds from the assumption that humanity will have brought upon itself a half-metre rise in the world's sea levels by the year 2100, even if all additional trace gas emissions into the atmosphere cease by the year 2030. Among the consequences would be 'flood catastrophes and the loss of homes for many millions', the Commission wrote. Why then does it not recommend the immediate launch of solar energy utilization on a broad basis, opening the door for even more drastic relief for the atmosphere long before the year 2030?

The question becomes more urgent because there is no way to guarantee a successful and practical realization of the potential for energy saving. While nearly everybody has emphasized the need for energy conservation in recent years, worldwide energy consumption continues to rise. Thus the German Parliamentary Commission notes in its 1992 report that, beginning in 1987, the commonly quoted, relatively modest reduction goals of 5% to the year 2005 would now require a decrease of 12% to reach the same quantitative goal, simply because CO_2 emissions rose another 7% between 1987 and 1990.[70] Not even this depressing observation has led to any reconsideration of these recommendations, nor indeed to the development of an up-to-date solar energy strategy to be implemented with full force.

The following structural reasons argue against the hoped-for and necessary success of an 'efficiency revolution':

1. Ever since the Industrial Revolution there has been a continuous increase in energy efficiency. The rapidity of this process differed, however: it moved faster in times of expensive energy and tougher industrial competition, but slowed down in times of cheap oil or an oil glut. The oil crises of 1973/74 and 1979/81 triggered the biggest jump in efficiency so far, resulting in the uncoupling of economic growth and energy consumption in many Western industrial countries. But the fact that the countries with the highest energy efficiency, the Western industrial countries, which account for some 20% of the world's population, consume about 70% of the world's energy supply each year clearly indicates that higher efficiency in energy conversion technology does not automatically mean overall energy savings. If there were such a direct linkage, energy consumption should occur at exactly the inverse ratio: the industrial countries, with 20% of the world's population, should be consuming less than 20% of the total energy. Those countries with lower energy efficiencies and 80% of the world's population should consume more energy than their global population share. In other words, it is impossible to talk about the effect of energy conservation independent of economic and social structures.

Higher energy efficiency reduces energy costs after a short period of time and brings with it competitive advantages. This cost reduction provides entrepreneurs, individuals and national economies with more freedom to manoeuvre to meet their energy demand for additional services or products. Savings in one sector of energy technology stimulate the acquisition of other energy technologies and more usage – the second or third car, more light fixtures or domestic appliances, more travel, but most of all, increased production. The energy savings concepts currently in vogue in the industrialized world are based on the unspoken assumption that one's own energy-based demand for services and products is largely met. This means that any improvements in efficiency affecting this demand would lead to a correspondingly lower demand for primary energy.

This assumption is not tenable in terms of the sociology of economics. Developed industrial societies are invariably the greediest for additional energy services. In order to achieve effective energy savings in the economy via increased energy efficiency in companies, simultaneous increases in energy costs must be guaranteed. Efficiency improvements that merely lower costs lead inevitably to increased consumption. Tax increases must kick-start energy savings, and further tax increases must be instituted to catch and contain cost reductions, which can otherwise cause reckless growth in energy consumption. Only if this type of political rationality underlies the industrialized countries' strategies towards increased efficiency can we hope to achieve genuine energy savings in the overall economy.

2. Conventional wisdom is that particularly the former communist countries and the developing countries offer great potential for improving energy efficiency. Statements of this type usually refer to isolated technical rather than sociological structures. This is demonstrated by the examples of the former Soviet Union and China. In industrial production, in power generation engineering, in the use of engines and equipment and in domestic and control techniques, both countries clearly employ less efficient energy technologies. Nevertheless, in 1990 the Soviet Union, with 40 million more inhabitants than the United States, emitted one billion tons of CO_2 less than the United States, and China, with more than five times the population, emitted 2.7 billion tons less than the United States. Despite greater energy waste, due to inferior technology and the system of state planning, the Soviet

Union and, even more so, China were operating at a lower overall level of energy use than the West. The reason is that their consumption level is less developed – in China, because of the absence of private cars and, by and large, exemplary utilization of organic agricultural wastes.

Suppose for a moment that the Soviet Union and China had operated during the last 40 years under the same type of economic system that exists in the United States, Western Europe and Japan: the storms and surges of a climate catastrophe would probably have demolished everything by now. If a Western-style market economy is to be introduced in these countries, then the history of the development of energy systems teaches us that two things will happen simultaneously: on the one hand an increase in economic energy efficiency as part of a general improvement in productivity, and, on the other hand, a clear general increase in energy consumption. This is not a contradiction, but the experience of 200 years of industrial and market-economic development. In most developing countries, the demand for energy-consuming appliances is almost entirely unmet.

The inability to distinguish between the structures of developed industrial countries and those of developing countries is evident even in respected scientific institutes. A 1990 study by the World Resources Institute entitled *What the North Can Do* argues against low energy prices for developing countries because they would encourage economic stagnation and would leave untapped the 'enormous savings potential'.[71] In other words, an analytical experience that is quite correct for industrialized countries is transferred in its entirety to the socio-economic conditions of developing countries. However, that 'savings potential' for billions of undernourished human beings translates into even more widespread hunger, more destruction of vegetation, and less agricultural and industrial production. The developing countries are not choking on their indebtedness because of low prices on world energy markets. On the contrary, energy prices are too high in relation to their level of economic development and their balance of payments. There is no doubt that the developing countries need more, cheaper energy for their economic development, just as the industrial West won its superior economic position via its advantages in the availability of energy, including a price advantage.

While one has to regard the efficiency revolution as a means to reduce the use of energy in industrial countries, always assuming that higher taxes can then prevent an increase in energy consumption, the increase in efficiency in developing countries is primarily to overcome the lack of energy. For social reasons, an increase in energy supply, efficiently used, is urgently needed in developing countries. To meet this demand with traditional, environmentally destructive energy has become impossible because of the lack of purchasing power on world energy markets (both for primary energy and for energy technology) and for environmental reasons. Attempts must be made, therefore, to make solar energy available to developing countries as soon as possible, in parallel with efforts to increase efficiency.

3. Among the preconditions for success in energy savings is a high level of awareness and information on the part of the people, well-trained technicians and a functioning public administration that can take the necessary steps and enforce them via regulatory activities – advice, heat and thermal regulations, key energy data for buildings, cogeneration systems, technical norms and standards, and many other activities. However, even in countries with relatively high levels of information and education, there are still striking shortcomings, both among the general public and among technical personnel. The vast majority of these countries suffer from a badly functioning public administration where rules and regulations frequently exist only on paper. For this reason it is unrealistic to expect, as was stated by the German Parliamentary Commission 'Protecting the Earth's Atmosphere', that 'as many states as possible, especially the United States, the European Community, Japan, the Soviet Union, the People's Republic of China, India and Canada, determine their reduction goals in the energy sector with a comparable methodology'. The United States, because of its sociological and administrative condition, would be able to do that in some regions, but not in others. It seems feasible in the European Union, by and large, and in Canada, Japan

and the Scandinavian countries. But what happens in the states of the former Soviet Union, the Philippines, in Tanzania, Brazil or India? To expect the same kind of administrative implementation of political energy savings measures as in the industrial countries is to fail to recognize national realities.

The sociological experience is such that energy sources – as long as they are geographically and financially available – are used without much thought until they run out. The exploitation of vegetation in the Third World will continue to the bitter end, as was the case in today's industrialized countries before the import of energy from abroad. It is difficult to see how that can change, unless there are functioning administrative or financial forces in place that would prove otherwise. The limited value of information and energy consciousness is demonstrated in the rich countries, whose inhabitants are more enlightened about the consequences but who nevertheless do not reduce consumption unless economic or administrative constraints force them to do so.

Even in the industrialized countries it is becoming increasingly difficult to expand the means of control and standardization needed to enforce energy savings and environmental mitigation successfully. The volume of environmental law is growing. Legislation and legislatively mandated action are becoming increasingly complicated the more areas they have to cover: clean air, clean soil and water, waste removal, nuclear safety controls or construction standards. Even countries with well-functioning administrative systems are finding it different to enforce environmental protection rules. The general cost of administration and the costs of safety and environmental regulation have not yet been taken sufficiently into consideration.

Based on its maximum achievable savings potential, a strategy of energy savings alone, therefore, is hardly sufficient to conduct an effective policy to protect the global environment. The frequently quoted sentence 'Energy saving is the largest new source of energy' is factually wrong because saving energy is not a source but merely reduces the demand placed on sources already being exploited. Additionally, it distracts from the fact that solar energy is the only truly alternative and environmentally friendly source of energy. If 50% of conventional energy can be saved, emissions caused by the other 50% will remain. A success rate of 50% in energy savings simply means that the emissions generated in one year at the moment will in future take two years to generate. Let there be no misunderstanding: the intent here is not to put a damper on energy-saving initiatives. All these efforts are needed, without any ifs or buts. The greater the success rate in energy savings, the faster solar energy can be used, because the total amount of traditional energy to be substituted by solar energy will be that much smaller. Energy saving represents a bridge to solar energy, but not a substitute. Both strategic elements belong together, but solar energy must be at the core of a strategy; solar energy alone is the definitive solution. Energy saving represents the vehicle to reach the goal more quickly.

Conflicting Perspectives: Solar Energy or Nuclear Fusion on Earth

For decades there was little controversy about the peaceful use of nuclear energy. The leading industrial nations spent several hundred billion dollars of public funds for the development of nuclear reactors, and they subsidized the construction of the first commercial reactors. Nuclear research centres were established, and the study of nuclear physics was the centre of interest for a generation of physicists from the 1950s to the 1970s.

Beginning in the mid-1970s, the consensus about the peaceful use of nuclear energy broke down. The public became conscious of the danger of nuclear disaster, as well as of the future liability of nuclear waste. The link between centralized power supply and environmentally ruthless economic growth became a theme of public debate. During the 1980s, nuclear energy became by far the most controversial form of energy. The resistance to atomic technology reactivated the environmental movement in the Western democracies. The movement in turn collided with plans of the Western industrial countries, which had begun to forge ahead with the expansion of nuclear power after the 1973/74 oil crisis, in order to become more independent of oil. Popular referenda in Austria and Italy halted the use of nuclear power. Sweden decided on a plan to abandon nuclear power altogether. In the United States and Germany, expansion plans for nuclear energy froze, although without any formal decisions being made. Not a single new nuclear power plant has been built in the United States since 1973. Even before the Chernobyl reactor accident confirmed the dangers of a nuclear accident, which until then had been regarded as extremely unlikely by the supporters of nuclear power, the euphoria of nuclear power plant producers, who had expected a massive worldwide expansion, had been dampened by the cold shower of a new disillusionment.

All this occurred before the start of the worldwide debate about the danger to our climate posed by the combustion of fossil energy, a debate that became unstoppable after the 1988 World Climate Conference in Toronto. Ever since, the supporters of nuclear energy have been hoping for a nuclear renaissance that, despite all their efforts, has not yet occurred. In self-pitying full-page newspaper advertisements in 1992, large German utilities quoted Albert Einstein's dictum 'It is easier to break down an atom than a prejudice'. They also argued:

> Let's be realistic: the more human beings inhabit the planet, the more energy and electric power is needed. Natural sources of energy such as water, sun and wind alone cannot meet the demand. In Germany, for instance, their share amounts to only 4% of the total electric power supply, despite large efforts. We must rely, therefore, on electricity from coal and nuclear power. Especially on nuclear power, because we must reduce the emission of carbon dioxide into the earth's atmosphere.

But it was not only the 'prejudices' of the environmental movement that stopped the worldwide nuclear offensive in its tracks. Not even nuclear supporters in governments and corporations have been able to ignore the fact that to date there is no safe method for disposing of nuclear waste. Above all, the dangers of nuclear weapons proliferation via the civilian use of nuclear power have turned out to be much larger than had long been admitted. The operational safety of nuclear power plants also turned out to be lower than originally assumed, an aspect that affects the use of nuclear power at its economically most sensitive point, because nuclear power is intended primarily for continuous base-load supply. Cost increases occurred which put drastic dampers on the willingness to invest, not only because of the extra safety requirements but also because of the high decommissioning costs. Privatization of Britain's utilities during the Thatcher era initially failed because private investors were not willing to buy into the risks of nuclear power plants, which are, ostensibly, the least expensive method of power generation!

Alarming reactor breakdowns such as Harrisburg in 1978 and the Chernobyl disaster demonstrated that an infrastructure to protect against catastrophes was required, something that exists in only a few countries. In developing countries, expansion plans were largely stopped because it became increasingly evident that large power plants are not suitable for their supply structures. Not only are there no long-distance transmission grids, but it would also be economic nonsense to build such wide-ranging grids for what is mostly decentralized demand. For that reason, even the so-called 1989 World Energy Conference in Montreal, supported by, among others, the producers of nuclear power plants, found that nuclear power utilization is not much of a prospect for the energy supply of developing countries. It is amazing how long it took the established experts to come to this obvious conclusion. If developing countries still display a special interest in nuclear power technology, it suggests a military rather than a civilian motive. Alternatively, it is prompted by their own energy experts, who are, of course, trained by and oriented towards the Western industrial model.

A companion problem is that uranium deposits for the production of nuclear fuel rods are also limited, making the perspectives for the use of nuclear energy even more problematic. Nuclear energy generates about 18% of the world's electric power supply from 420 reactors of different sizes. If nuclear power were to replace the output of all currently operating fossil-fuel power plants, an additional 1,000 reactors with an average output of 1,000 MW each would be needed, according to calculations by the Forum Atomique

Européen. If the total demand for heat, including industrial process heat, were to be met by nuclear power, a further 1,000 such reactors would have to be added to the grid; and if we were to replace the fuel needed for road vehicles with nuclear electricity, yet another 1,200 nuclear power plants would have to be installed.[72] Assuming a 30-year service life, providing the 3,500 large reactors needed for the global replacement of fossil energy would mean that a new reactor would have to be built every six days. It would also mean that every 30 years we would be faced with 3,500 radioactive reactor 'ruins', and furthermore, if the world energy demand continues to rise, the total number of new reactors would have to be significantly increased as well.

This would result in more rapid depletion of the world's uranium deposits, much greater potential for accidents, and much more severe problems of waste disposal. The initial approach to the problem was to develop technologies that would 'stretch' fissionable materials, by reprocessing nuclear fuels and the development of fast breeders. These technologies are increasingly fragile and subject to breakdown, and would be more expensive than normal nuclear power plants. The German fast breeder in Kalkar, which was supposed to cost $125 million, was cancelled in 1991 before start-up after spending reached $4.4 billion. The French Super-Phénix fast breeder started operations in 1986 but operated for less than 200 days before it was shut down in 1992 after constant breakdowns – a veritable money-shredding machine. At the same time, the danger of using the plutonium produced by fast breeders for military purposes increases rapidly. Since it is evident that no government will rely on nuclear power without access to fast-breeder or reprocessing technologies, because of the limited nature of uranium deposits, nuclear power itself presents a stumbling block to the goal of preventing the proliferation of nuclear weapons. It is also clear that the control of the nuclear cycle demands a large measure of administrative reliability and political stability within a state, which means that any increase in political instability once more intensifies the dangers of nuclear power.

There must be deeper, more profound reasons why governments and energy enterprises of the industrial world continue to cling to nuclear technology. It is true that most of them talk about nuclear power overwhelmingly in terms of 'transitional technology', without being very specific as to where this transition will lead. Some say it will lead to smaller, 'decentralized' nuclear power plants: not the 500, 1,000 or 1,500 MW ones of the past, but 100 or 200 MW capacity per plant perhaps. This overcomes none of the bottlenecks mentioned earlier. Rather, the hoped-for solution has to do more with the perspectives opened up by nuclear fusion. This approach is supposed to fulfil at some point in the future the promise of a lasting, permanent energy supply that is obviously impossible to realize with nuclear fission. Nuclear fusion appears as the culmination of the path taken with nuclear energy, a historical justification that in principle the path taken since the 1950s has been the correct one after all.

Fusion technology is a special variant of nuclear power. Instead of splitting heavy atomic nuclei to produce power, fusion, as the term implies, fuses lighter atomic nuclei into heavier ones, producing an energy surplus that could be used to generate heat and electric power. Many of its advocates try to deflect attention by pretending that atomic fusion is something quite different. When protest against nuclear fission reactors was at its height in the mid-1970s, nuclear fusion researchers dropped the word 'nuclear' and called themselves just fusion researchers from then on. In so doing they managed to shelter from the worst of the criticism of nuclear energy. However, the physical difference between a nuclear fission and a nuclear fusion reactor is the same as that between an atom bomb and a hydrogen bomb. In essence, nuclear fusion represents the attempt to copy here on earth the process of generating solar energy inside our chief star, the sun, 150 million km (94 million miles) from earth. 'Why should it be impossible to achieve on earth what functions in so many places in space?' asks the German fusion researcher Eckehard Rebhan in his book *Hotter than the Fire from the Sun*.[73] The current generally accepted date for the availability of commercial fusion reactors is around the year 2050. Towards that goal, the European Union and its member states alone are to spend some $62.5 billion in public funds between now and the year 2050. Some $3.1 billion were spent between 1950 and 1976. After that, the trend zoomed steeply upwards: between 1976 and 1987 another $3.2 billion went to fusion, and between 1987 and 1990 the total was $2.9

billion; in other words, about $938 million annually. The goal for the period 1991–2010 is $12.5 billion, or about $625 million annually; between 2010 and 2025 $18.8 billion, or almost $1.25 billion per year; and another $12.5 billion after 2025.[74] It would be very surprising if these advance estimates do not rise significantly if the programme continues. The first joint European test reactor, JET, cost a total of $1.25 billion; its successor, NET, is already expected to cost eight times as much.

Table 6 *R&D budgets of OECD member countries (in millions of $ at 1995 prices)*

	Nuclear fusion	Renewable energies
1984	1,523.8	1,098.2
1985	1,509.8	910.6
1986	1,390.2	711.8
1987	1,315.3	671
1988	1,208.1	676.8
1989	1,138.2	619.3
1990	1,163	664.8
1991	1,064.4	697.3
1992	1,035.5	766.3
1993	1,121.6	774.7
1994	1,073.5	790.2
1995	1,105.1	896.8

Source: OECD/IEA, *Energy Policies of the IEA Countries*, 1996 Review

Table 6 shows the research budgets for nuclear energy and fusion energy, compared with those for renewable energies. Once again the table shows a grotesque imbalance, re-confirming the evaluation of the 'Faustian Age' made in the second chapter. Moreover, the table does not include the large sums that the EU Commission has made available year after year for nuclear fusion research, compared with the relatively paltry amounts for renewable energies. It can only be irresponsible amnesia combined with blind faith in technology obsessed with science fiction that gives precedence to unrealistic nuclear fusion over realistically useful solar power.

Proponents cite the inexhaustible supply of fuel as fusion's overriding advantage. (Deuterium and tritium nuclei are fused in the process. Deuterium, the argument goes, is available in practically inexhaustible quantities in the world's oceans, and tritium is made from lithium, which is also available in near-limitless amounts in the oceans.) They also cite its high degree of environmental compatibility; safety, assured because the kind of runaway meltdown much feared with fission reactors is impossible; and only a slight potential for biological danger. The fact that fusion needs such a long, expensive time for its development – and it is still questionable whether there will ever be a commercially operable fusion reactor – is the result of a whole series of massive problems. Fusion requires high pressures and temperatures of more than 100 million °C in order to heat deuterium and helium to levels at which electric power can be derived from their plasma state. Materials capable of withstanding such temperatures for long periods of time still need to be developed. In any case the wall of such a reactor vessel will probably have to be changed at least every six weeks.

According to preliminary calculations, some 2000 m^3 of high-level radioactive material that will still be active after 100 years will be generated during the operating lifetime of a fusion reactor, plus some 4000 m^3 of medium-level radioactive material – overall, much more than in a conventional fission-type reactor, although the radioactivity would decline much more rapidly. The exchange of materials would have to be handled by robots. The plasma must be confined magnetically, requiring large quantities of electric energy to start the power-generating fusion process. This requires superconducting magnets that have to be cooled down to minus 270 °C:

> Only highly imaginative specialists have been able to figure out so far how to control safely in a tiny space, without accidents, plasma temperatures of 100 million degrees and temperatures of minus 270 degrees to maintain the huge current for the magnets under extreme vacuum conditions and with serious radioactive loads

is the way Professor Helmut Tributsch describes the problem.[75]

Still open is the question of what happens if the cooling system breaks down. Tritium produced inside the reactor can penetrate solid structures. In contact with air, it forms tritiated water, which, once introduced into the natural water cycle, can cause severe biological damage. Lithium is highly corrosive, and in contact with water, air or concrete produces extremely hot chemical fires. The volume of cooling water required for a fusion reactor is many times more than that needed for fission reactors, and can easily boil. Because of the need to change plasma containment walls frequently, long down-times would be the norm, requiring extra generating capacities.

These few remarks should suffice for now. They cannot replace a precise, serious debate about the particular aspects and problems of fusion technology, but a basic debate in terms of economics and energy policy is possible in an abbreviated discussion employing a 'best case', rather than a 'worst case', scenario. 'Worst case' assumptions that used to characterize the conflicts about nuclear weapons or nuclear power plants proceeded from the worst imaginable danger – a nuclear war or a Maximum Credible Accident. The main question about fusion energy is what dangers realistically crop up through copying on earth the energy release process that occurs in the centre of the sun. A reasonable assumption would be that fusion technology, put into practice, would overtax human abilities as much as nuclear fission technology. A best-case assumption, on the other hand, proceeds from the standpoint of the optimistic fusion supporters that all the problems hinted at are manageable in a responsible fashion and are, in any event, decidedly less severe than those encountered in energy produced from nuclear fission. Even proceeding from a best-case assumption in discussing fusion, so many massive arguments against it remain that they lead to a clear rejection of the idea of continuing the project.

1. Until now it was regarded as almost naive to ask about the costs of a grandiose technology effort such as fusion energy. Some estimates, however, predict energy generation costs roughly ten times as large as those associated with fission reactors.[76] Fusion power plants will be very expensive because the surface needed to exchange heat must be very large. Additionally, the 'energy payback time' of a fusion reactor – the period of time needed to produce more energy than required to construct it in the first place – is likely to be uneconomic. The physicist Benecke estimates that period to

be 20 years for a reactor with a lifetime of probably only 30 years.[77]

But assume the most optimistic position, which still admits that fusion-generated electricity would cost two to three times as much as that generated by conventional fission reactors.[78] By way of comparison, it should be noted that there are even today – at the very start of the utilization of solar energy! – wind power plants, solar power plants or biomass plants that match the operating costs of existing nuclear plants or that produce power at lower cost, even without including the cost of nuclear waste disposal. Even if there were not a hint of danger associated with the operation of fusion energy, there would be no economic reason to pursue it, as electricity from solar power is already more cost-efficient than can be expected of fusion in the best possible case. If fusion energy ever becomes commercial it is safe to predict that solar energy will be far more economical. There is not the slightest doubt that solar energies will continue to become cheaper with increasing usage. This means that only if the potential for solar energy is not large enough to assure humanity's energy supply would there be a reason for continuing to support fusion energy. The potential is in fact inexhaustible, global and much larger than will ever be needed. In short, fusion energy is superfluous in view of the real opportunities for solar energy. Even if there were no problems in managing fusion, mankind simply does not need such a form of energy.

2. A fusion reactor would be an even more extreme example of centralized, expensive technology than a conventional nuclear power plant. A conventional reactor with a 1200 MW block contains a core weighing 1,000 tons; the core of a fusion reactor would range from 5,000 to 30,000 tons. A fusion reactor would have an output capacity of 5,000 MW or more – there are already design sketches for 200,000 MW behemoths – which by comparison almost gives a conventional nuclear power plant the aura of a decentralized plant. Fusion energy requires exactly the opposite structure to that of technical solar energy use, which is decentralized. The prospect that the existing, mostly centralized power supply structure would reach a

climax of sorts with fusion energy is apparently one of the reasons why the power industry perseveres in its insistence on pursuing that line of development and pointedly ignores the obvious advantages of solar energy. With fusion energy on the scene, the existing energy supply structures, which for all practical purposes already represent states within the state, could be perpetuated indefinitely into the future. Artificial fusion and natural solar energy delineate two essentially opposing development philosophies. Those who are waiting for fusion energy have to maintain and expand the existing centralist management structures. Users of solar energy must strive for diametrically opposite concepts. If solar energy is used now and extended further and wider, an energy system will be created into which fusion reactors will no longer fit.

3. Fusion energy is without a doubt technologically even more complex than nuclear fission. Because of the extremely long technological lead time required, only North America, Japan, Western Europe and perhaps Russia would be able to master these technologies. Since fusion is regarded as the energy supply for the time after the end of the fossil fuel and fission era, control of this technology by only a few implies a lasting position of political and economic predominance over the rest of the world. Maybe this is another reason why the energy establishment continues to cling to the fusion idea, and wants to ride out the time between now and fusion's arrival with current energy carriers, bypassing solar energy.

4. Fusion energy can hardly be described as friendly to the climate. Since fusion generates additional heat in the earth's atmosphere, it is a category of energy carrier considered undesirable according to entropy law criteria. Even if these climatic pressures are not as easily measurable as trace gas emissions, they will have consequences for the global heat budget.

Although any comparison between the hypothetical prospects for fusion and the actual prospects for solar energy favour the latter, fusion remains the favourite for development efforts. There is no rational basis for that any more. Nevertheless, even official scientific institutions feign blindness when confronted with the solar energy issue. Take, for example, a passage at the end of a 450-page book published recently by a German fusion researcher:

> Perhaps the conversion of our energy supply system to solar energy would be the ideal solution. But at present there is nothing to indicate that this is a realistic option.[79]

To declare that fusion is realistic, but solar energy is unrealistic, has no connection any more with scientific opinion and does not deserve to be taken seriously, were it not for the fact that it is often a government-endorsed view.

This also holds true for the internationally renowned energy and social scientist Cesare Marchetti from the International Institute for Applied Systems Analysis in Laxenburg, Austria, who has published a number of scenarios about the world's energy supply up to the year 2100, in which fusion energy flowers profusely, but which don't have anything to say about solar energy![80] Solar energy is not mentioned because it is not supposed to be mentioned – that is the only explanation for analyses that carry the claim of scientific validity but which in reality are examples of hoodwinking.

The experts just quoted have since become outsiders, but even among the establishment experts the rule is still to apply different yardsticks when it comes to comparing fusion with solar energy. With great tactical skill, the promoters of nuclear fusion have been able to keep the subject out of any critical public debate of nuclear energy by describing it as a basic, less problematic alternative. There is no public debate about fusion energy, only occasional success stories. The experiment of 9 November 1991 in the JET experimental laboratory in Culham, England, was used to generate worldwide publicity for fusion. In that experiment, scientists generated 1.8 MWh of electricity for two seconds, not mentioning, of course, that it took 17 MWh of electrical energy to heat the plasma for the experiment. 'Another chapter was written on this day in the history of a hitherto successful advertising campaign in favour of nuclear fusion', commented one report,[81] right on time for the decision-making process of the Commission of the European Community to unfreeze funds for fusion research that had been blocked by the European Parliament. The Parliament had put a hold on the funds to try to force the Commission to support solar research to the same extent as fusion but this effort failed because of, among other reasons, the Culham experiment's dazzling effect.

'If we don't look out, the science managers and the politicians will sell us an alternative that isn't a real one', cautioned one report several years ago.[82] Even the warning of the former deputy director of the Plasma Fusion Center at the Massachusetts Institute of Technology, M.L. Lidsky, that 'If the fusion programme produces a reactor, no one will want it'[83] remained unheeded in the upper strata of the political decision-making centres. The claims of its proponents, wrote Jeremy Rifkin, 'are eerily reminiscent of those made by fission advocates two decades ago'.[84] For energy researcher Amory Lovins, all nuclear energies 'are the equivalent of using a chain saw to cut butter'. In other words: we don't need them.

4

The Potential for and
the Keys to a Solar Strategy

Comprehensive proof must be provided that the potential of solar energy is sufficient to replace conventional energy sources completely, that it can help to prevent climate catastrophes and other dangers deriving from conventional energy utilization, and that it is capable of overcoming the lack of energy in developing countries. The total solar energy potential is composed of the amount of energy actually available, the technical potential of energy conversion, and the economic potential for using these technologies. The remarks that follow are based on the current level of technical development, the cost of solar energy installations. and the cost of converted energy. They demonstrate a market of real, sweeping new possibilities.

There is no doubt that solar energy offers humanity a far greater energy potential than it will ever be able to use – inexhaustible as well as usable for all activities of all people, including industrial applications. At a distance of 150 million km (93.75 million miles), the sun radiates incessantly a mere fraction of its total energy output onto earth. In a quarter of an hour the sun offers more energy than humanity consumes in an entire year. Not all of it is, directly or indirectly, usable for man, but the usable potential is now one thousand times larger than mankind's total annual energy consumption (comparing total energy consumption with the total potential generating capacity using 15% efficient photovoltaic cells – the current state of the art).

In OECD countries alone, the usable solar harvest is 170 times the final energy consumption, as indicated in Table 7 on page 89. In the CIS countries and Eastern Europe the ratio is 400, while for developing countries it is 950. Other calculations indicate that solar plants with an efficiency of only 10% over a total area of 500,000 km^2 (193,000 square miles) in North Africa's Sahara desert would be sufficient to supply all humanity with solar energy. Added to direct solar radiation should be the potential for indirect solar energy: wind power, biomass, hydro power, wave energy. Only 20% of the annual growth of biomass is in theory sufficient to meet mankind's primary energy consumption.[85]

Table 7 *Solar insolation in relation to commercial energy demands (in billion tons of oil equivalent per year)*

	OECD countries	Eastern Europe and USSR	LDCs	Total
Yearly solar insolation converted into oil energy equivalent units at 15%	505	580	1,420	2,505
Commercial final energy consumption (1980)	3	1.5	1.5	6
Ratio energy consumption/ insolation at 15% average efficiency	1:170	1:390	1:950	1:420

Source: Dennis Anderson and Catherine D. Bird, 'Carbon accumulations and technical progress', manuscript, University College London and Balliol College, Oxford, December 1990, p. 20.

Scenarios for Active Solar Energy Utilization

Statistics differ widely in estimating the current share of solar energy in final energy consumption. With the exception of hydro power, the solar share is often ignored completely in these statistics. According to the statistical office of the European Union, renewable energies contributed 7% to the EU energy supply in 1996, if we base their substitution value on nuclear/fossil primary energy.[86]

Scientific scenarios assessing the possible share of solar energy in a nation's energy supply arrive at results that are far above those claimed by the notorious sceptics. A study by five American research institutes analysing the potential for the United States found that, by tripling research and development efforts, solar energy could achieve a share of 29% of the total American energy supply by 2030. Additional political stimuli (higher taxes for conventional energy, tax concessions on solar energy, and reduction of administrative and information-based barriers to the introduction of solar energy) could increase that share to 50% or more in the same period.[87]

In 1994 Eurosolar demonstrated, in a study for the European Union, what concrete steps could be taken in order to reach a 50% share in the EU by 2020. They picked the most dynamic developments in each segment of renewable energy in each of the EU countries as the basis of calculation. Achieving the desired result depends on the continuous further development of such initiatives – including photovoltaics, wind power, solar thermal energy use and biomass – and their widespread adoption in all the EU countries. In this calculation they did not include any assumptions about government stimuli, in order to avoid the concern that governments and economies would be overstretched.[88]

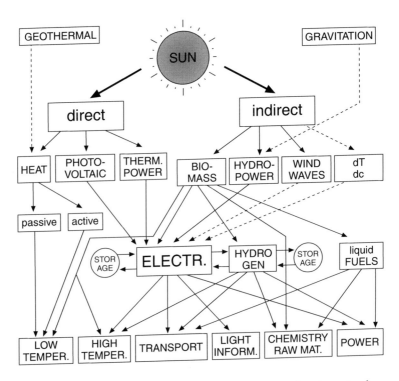

Figure 3 *The 100% renewable energy scenario for Austria, by Hans Schnitzer*
Source: *Yearbook of Renewable Energies 1995/1996*, p. 88

In the 1995/96 edition of the *Yearbook of Renewable Energies* various scenarios were documented that demonstrate how renewable energies can provide 100% of our energy requirement in the future.[89] Hans Schnitzer calculated a 100% scenario for Austria in 1990 at the Institute for Chemical Engineering at the University of Graz.[90] The main strands of this outline are shown in Figure 3.

As long ago as 1982, at the International Institute for Applied Systems Analysis (IIASA) in Laxenburg near Vienna, which is an

Table 8 *The current state of forms of energy in 1975 compared with the 'hard solar scenario' and the 'soft solar scenario' for 2000 of the International Institute for Applied Systems Analysis*

a *Final Energy by Form and Use, Base Year 1975 (GWyr/yr)*

Final Energy Form	Final Energy Use						
	Thermal Low	High	Coke	Feedstocks	Motor Fuels	Electricity	Total
Coal	48.7	28.8	43.5				121.0
Oil Products	335.9	52.6		81.2	239.9		709.6
Natural Gas	109.1	86.7					195.8
Electricity*						141.2	141.2
Biomass	26.4						26.4
Total	520.1	168.1	43.5	81.2	239.9	141.2	1194.0

* Electricity predominantly from nuclear or fossil fuel

b *Final Energy by Form and Use, Hard Solar Scenario, 2100 (GWyr/yr)*

Final Energy Form	Final Energy Use						
	Thermal Low	High	Steel Production	Feedstocks	Motor Fuels	Electricity	Total
Coal			2.0				2.0
Electricity**	68.2				31.6	644.0	743.8
Biomass	87.9						87.9
Methanol				518.0			518.0
Hydrogen	689.6	315.0	59.9		390.2		1454.7
Total	845.7	315.0	61.9	518.0	421.8	644.0	2806.4

** Electricity from renewable sources

c *Final Energy by Form and Use, Soft Solar Scenario, 2100 (GWyr/yr)*

Final Energy Form	Final Energy Use							
	Thermal Low	High	Steel Production	Feedstocks	Motor Fuels	Electricity	Cogeneration	Total
Coal			1.0					1.0
Electricity**	16.1	50.6			20.9	160.1		247.7
Biomass	13.8						96.2	110.0
Methanol				216.3				216.3
Hydrogen		1.7	31.2		173.5			206.4
District Heat	34.4							34.4
On-site	344.7	70.5				156.5		571.7
Total	409.0	122.8	32.2	216.3	194.4	316.6	96.2	1387.5

** Electricity from renewable sources

Source: *Yearbook of Renewable Energies 1995/1996*, p. 97

institute more oriented towards nuclear energy, two 100% scenarios for renewable energy for Western Europe were calculated with a target date of 2000. The first, a so-called 'hard solar' scenario, used a centralized energy system based on solar hydrogen as the main form of energy. The second, a 'soft solar' scenario, had decentralized energy sources based on meaningful increases in energy efficiency. Table 8 shows these two scenarios compared with the status quo in 1975.

Such scenarios were never disproved. They were simply ignored by the majority of the so-called experts as something that cannot be allowed, cannot be 'realistic'. After all, if it had been admitted that comprehensive solar energy provision were feasible, the attempt to continue with the nuclear/fossil energy system would have lost all legitimacy in the eyes of the public.

An especially interesting concept was presented in 1978 by the Groupe de Bellevue, a group of scientists previously mentioned on page 44. Taking into account the potentials of biomass, photovoltaics, solar thermal power plants, hydro and wind power plants on the basis of 1975 conversion technologies, this group calculated that France's energy needs could be met entirely by solar energy. Based on these types of solar energy, the group calculated supply shares of 11% for liquid fuels, 14% for solid fuels, 11% for gaseous fuels, 34% for heat and 30% for electricity, equal to 140 million tons of crude oil equivalent. The scientists based their estimates not on the energy demanded, but on the amounts of energy actually required – in other words, on

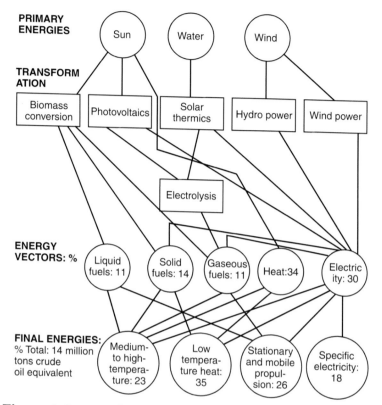

Figure 4 *Design of a completely solar energy supply system for France: long-term energy schematic*
Source: Groupe de Bellevue, 1978

a basis of systematic energy-saving measures – for 60 million French people, assuming a high standard of living for all of them. In their scenario, solar heat provided 80% of the total needs for buildings and 40% of industrial heat, with land requirements for collectors totalling 250,000 ha (2,500 km²). Electricity would be generated from hydro power, tidal power, wind and solar power plants, which also produce a major share of the hydrogen needed under this concept, requiring 450,000 ha (4,500 km²). Liquid fuels and gas were to be derived from biomass, requiring 5 million ha (50,000 km²) land area of constantly afforested woods (one third of France's total woodland) and 2.5 million ha (25,000 km²) of agricultural land, out of a total of 35 million ha currently under cultivation (see Figure 4). This study did not attempt any cost calculations because the goal was merely to demonstrate the technical feasibility. The potential has expanded considerably since then, making it even easier now to meet the requirements of a total, solar-based energy supply. The individual elements would be different: notably, less land would be needed. However, even the land demand calculated for 1978 shows that it would be easy to integrate such a system into existing agricultural structures and buildings.

The validity of the information offered in scientific scenarios of this type is by definition limited, because they have to proceed from currently provable numbers, and because they cannot assume any development breakthroughs or great leaps forward, or allow for the influence of special interests, crises and political changes. However, it can be safely assumed that the current application potential of these mostly young technologies for renewable energy utilization can only improve, both technically and economically. There are no negative surprises, only positive surprises in the pipeline. Since nature's energy store is large enough to meet all our energy needs with solar energy, there is no reason not to try to reach 100% coverage, given the remaining danger potentials. The frequently asked question of how large the share of solar energy could be in the total energy picture is actually foolish: since solar's potential is more than sufficient to meet human energy needs, there is no limit to the usable solar energy share. The order of magnitude of the solar energy share is simply a question of 'input': the more political effort and economic investment, the larger the share.

That these potentials exist will be demonstrated – not with another scenario, but with a roughly sketched outline that employs selective extrapolations of the various approaches to the use of solar energy. This is based on an irrefutable assumption: if it is possible to manufacture 50 MW worth of solar cells or 1,000 wind power plants per year, it must also be possible to manufacture 50,000 MW of solar cells or 1 million wind power plants or more every year, since there are no limits in terms of raw material availability. Convincing proof is provided by the production of cars, which are, after all, simply decentralized energy converters. The more than 4 million cars produced every year in Germany alone represent an energy conversion capacity of more than 200,000 MW, roughly double the total current German power plant capacity. This calculation is based on assuming an average mechanical output of 50 kW per passenger car. In 1989 there were 128 million passenger cars in the member states of the European Community with a total energy conversion capacity of 6.4 million MW. In stark contrast, there existed only 411,000 MW in generating capacity on the part of the electric utilities. Whatever a modern industrial society is capable of achieving in terms of automobile output, it should be capable of producing by solar energy technology as well. In other words, the only really important aspects here are the amount of capital that needs to be mobilized, and the power of political organization and persuasion.

These cost studies should not be overemphasized. Even if solar energy's 'competitiveness' simply isn't there in comparison with that of conventional energy carriers, and could never hope to match it, the path towards solar energy should be taken, without hesitation, in any event because of the overriding concerns for the continued assurance of human existence. For the same reasons, there were seldom any questions about costs when it came to matters of military security, unless it was a matter of comparing the costs of different weapons technologies to achieve a given goal. With solar energy, the question about costs or the 'energy balance' is raised with such notorious frequency, because the questioners typically want to use the answers as an argument to block solar energy. High costs are supposed to have an intimidating effect and to prove economic 'unreasonableness' or even 'irresponsibility'. Enormous costs did not matter at all when nuclear power was at the early stages of development, because the goal of a supposedly inexhaustible energy supply in the future was paramount. These comparisons show the carefully planned poisoning of the atmosphere surrounding solar technology, and how strenuously the opponents were searching for killer arguments against solar – as if human beings operate at every instant of their everyday lives exclusively according to 'economic' principles. However, it is obviously true that the conversion of our energy system into a solar energy system will be achieved more rapidly and with less friction the lower the costs.

Potentials That Can Be Realized Immediately

A basic precondition for a successful solar strategy is that not just one option for using solar energy be tried, but that all relevant options – from photovoltaics to solar thermal power, from wind power to biomass, from hydro to hydrogen, all based on the optimal level of efficiency – will be tried within an overall programme. This truism has to be mentioned again because one option is frequently played off against another, especially energy conservation against solar technology, even by supporters of environmental politics. A serious and responsible strategy aiming genuinely at replacing the current energy system must leave such petty controversies behind. This includes no longer thinking traditionally in terms of successive stages – energy saving first and only then solar energy, for example – but of advancing in parallel steps. As has been made clear, it is by no means true that energy saving is always the prerequisite first step – especially if new investments for energy facilities are at stake.

Energy saving obviously remains a key element on the road to a solar energy economy. Energy-saving strategies will not be discussed here to any great extent because their importance is no longer seriously in dispute, and because an extensive body of literature on the subject exists already. However, the frequently cited advantage – that they represent the quickest and most effective method for climate relief – is vastly outperformed by another strategy: the carefully targeted utilization of solar energy by nature itself.

Nature's Key to the Sun: Photosynthesis

Since plants require carbon dioxide for their growth, and bind it as carbon via photosynthesis, they cause an exchange of carbon between the atmosphere and the biosphere. They are repositories for carbon or, rather, CO_2. If trees are not merely burned but turned into wood products instead, carbon dioxide remains stored. Some 800 billion tonnes of carbon are thus bound into the globe's botanical system, while the atmosphere contains more than 700 billion tonnes.[91] Every year, almost 8 billion tonnes of carbon are released by the burning of fossil energy and by deforestation, which generates about 29 billion tonnes of CO_2 emissions. About three fifths of that total is absorbed by the world's forests and by the oceans (which might systematically supersaturate the oceans). About two fifths – and perhaps increasingly more – remains as additional CO_2 in the atmosphere, and this represents the biggest of all dangers to the climate. What more obvious solution could there be than a massive reafforestation effort on a global scale in order to remove the CO_2 that current forest stocks can no longer absorb from the atmosphere?

The reafforestation of 1 ha (about 2.5 acres) of forest binds at least 10 tonnes of CO_2 annually during the tree's growth, lasting about 40 years – or a total of about 400 tonnes.[92] Ten tons of CO_2 are about equal to the average annual amount of emissions generated by one person in the industrial countries. The costs of such reafforestation range from less than $65 per hectare in some developing countries to up to $12,500 in industrial countries, depending on the cost of labour and geographical conditions. These figures are based on the practical experiences of the American organization Global Releaf, which conducts such reafforestation efforts with private donations. Since it doesn't matter for the atmosphere where on the globe reafforestation takes place, the cheapest places can be chosen – in developing countries, where most of the free spaces, available for reafforestation, happen to be located. In those areas, reafforestation costs are lower by ratios of 1:100 to 1:200 compared with the cost of CO_2 reduction via technical measures in industrial countries. This approach would have other advantages as well: it would create numerous new jobs, improve the natural water cycle, help avoid soil erosion, improve the regional climate, and provide opportunities for building or expanding a commercial timber industry.

The planet's land areas that are free from ice total about 130 million km² (50.2 million square miles). Of that, slightly less than 40 million km² (15.4 million square miles) are forests, or almost 30%.[93] R.A. Houghton and G.M. Woodell, writing in 1989 in Scientific American, calculated for tropical countries alone an area of 8.5 million km² as suitable for reafforestation, and they even estimate that one hectare can bind 20 tons of CO_2.[94]

Frank Rosillo Calle and David O. Hall cite calculations according to which 17 billion tons of CO_2 can be captured by 7 million km² (2.7 million square miles) of reafforested land, or even as much as 11 billion tons by 3 million km² (1.16 million square miles) of reafforestation.[95] The Beijer Institute in Stockholm says that 5 million km² are enough to, in effect, relieve us of all CO_2-related problems.[96]

In order to make an effective contribution towards fending off dangerous climate changes with reafforestation, such an initiative must be planned on the grand scale. The creation of large-scale CO_2 sinks, some fear, could be abused to continue excessive consumption of energy. The same argument could be used, however, against energy conservation strategies. In fact, both reafforestation and energy conservation strategies have a bridge-building function designed to lead us with less risk into the new age. During the decades in which new forests will be growing, the planetary energy system must be converted into a planetary solar energy economy. Global reafforestation programmes represent a one-off act of strength, simply because the area suitable for reafforestation cannot be enlarged. It is incomprehensible why this key element of a solar strategy is not mentioned at all, or plays only a minor role in most debate. Reafforestation is no substitute for the utilization of solar power with the help of solar technologies, or for energy conservation measures, but is an indispensable programme for right now. 'Let us plant survival': this slogan, coined by Richard St Barbe Baker,[97] a Briton who died in 1982 and who planted several million trees on all continents during his lifetime, is more relevant today than ever before.

Technical Keys to Solar Energy

Industrial mass production of solar energy technologies promises cost reductions that compare very well with those of traditional energies. Dennis Anderson and Catherine D. Bird, writing in the May 1992 issue of the *Oxford Bulletin of Economics and Statistics*, have compared cost trends of traditional electric energy carriers between 1892 and 1980 in the United States with the average costs of solar thermally or photovoltaically generated power, as they have evolved since 1980 and projected to the year 2030: see Table 9.[98] These cost developments were then correlated with increases in production rates for solar technology and their utilization in areas with different insolation rates: see Table 10.[99] Referenced to individual solar power plants, another American calculation from 1990, based on market interest rates (in

Table 9 *Traditional sources of power generation in the USA, 1892–1980*

Year	1892	1900	1925	1950	1980
Real price, 1990 (cents/kwh)	230	140	40	9	6

Renewable sources of electricity generation (high insolation areas)

Year	1980	1990	2000	2020	2050
	40	18	13	7	7

Table 10

Capacity (in million kw)	0.8	5.0	36.0	300
Weighted average costs (per kw)				
high insolation areas	19	13.5	9.8	7.6
medium insolation areas	22.8	15.5	12.2	9.3
low insolation areas	31.9	21.5	16.5	12.9

Table 11

	1980	1988	2000	2030
Wind power	32	8	5	3
Photovoltaics	339	30	10	4
Solar thermal power with gas back-up	24	8	6	–
Solar thermal power with parabolic receiver	85	16	8	5
Electricity from biomass	5	5	–	–

cents/kWh), provides the data listed in Table 11.[100]

Even if the trend forecast for photovoltaics is too optimistic in terms of number of years, because the costs of 10 cents/kWh were not achieved by 2000, there is still little doubt that this figure will be achieved at a later date. In Japan the government and the research institutes calculated that when the 100,000 rooftop programme, which started in 1996, is complete, photovoltaic electricity generation costs would fall to 19–22 cents/kWh. That could therefore be achieved in the early decades of this century.[101]

These cost calculations make it clear that wind power, biomass and solar thermal electricity generation have nearly reached the cost levels of conventional energy carriers. Estimates for future cost trends of solar energies can be calculated much more reliably than can those for conventional energy carriers: the changes and trends depend almost exclusively on the production costs for solar technology, and not on the unforeseeable cost fluctuations for primary energy on the energy markets. Solar radiation and wind are free of charge, and will stay that way; on the other hand operators of biogas facilities may in fact be able to charge a fee and biomass from plants is not free of charge, because it will have to be cultivated commercially.

An examination of the various key elements provides even clearer indications of how the potential can be increased via the assembly line production of solar energy facilities.[102]

Solar Thermal Heat and Solar Construction

Most advanced and most widely used – especially in Israel, Turkey and Greece – are solar collectors that generate heat and hot water for domestic use. A 1992 study calculated for Great Britain that a collector area of 5–6 m² (54–65 ft²) produces 225 litres (59.4 US gallons) of warm water per day. Internationally, prices range from about £100/m² in Israel to £400/m² in Italy, depending in part on production volumes. If collectors of this type were installed in all private single-family homes in the UK, the result would be a CO_2 reduction of about 6 million tons.[103] Nothing stands in the way of their immediate introduction, because the acquisition and installation costs are matched, or even surpassed, by the cost of energy saved. Centralized solar facilities for space heating and warm water – already available in Sweden, for instance, for apartment complexes of up to 500 units and for commercial buildings – are approaching prices of no more than £150/m² of collector surface, as reported by another British study.[104] A cost projection of these data demonstrates that CO_2 emissions would be reduced by another 5 million tons for Great Britain. But all this is only a small beginning.

It has been proven that the passive use of solar energy in buildings – by appropriate alignment of the design to the angle of incoming solar light and its efficient utilization, with new facades and construction materials for heat accumulation and for cooling systems, with heat exchange systems, with utilization of the glass surfaces for power generation and via solar collectors – can save 90–100% of a building's heating costs.[105] This is limited to new construction, however, and requires an appropriate building site. While these new technologies produce some additional costs, these are compensated, sooner or later, by savings in energy costs. The house itself must be regarded in its totality as a solar collector and light directing system. There are multiple solar house building techniques: from daylight systems to solar cooling, from solar construction elements of all kinds for electricity generation to a combination of central heating furnace and solar collectors, and many more. Solar construction is an Eldorado (for creative architects).

Buildings have an average lifetime of about 100 years. It is conceivable that the entire heating requirement for buildings alone, which accounts for 25–40% of total energy demand in most countries, could be met totally via passive and active solar energy use. This would also have many additional beneficial effects, such as raising the general level of health in densely populated metropolitan areas and reducing residential utility costs (which in some European countries in effect amount to a 'second rent'), and thereby providing an indirect increase of income. Solar buildings feature a high degree of transparency, and in almost ideal fashion they combine individual personal independence with societal responsibility in an architecture for the new millennium.

It cannot be seriously disputed that photovoltaics can cover the electricity needs of single-family and even multiple-family dwellings. If batteries are not used for storing electricity, the production of the amount of electricity needed in one's own home, or even more, can be assured through continued parallel access to the power grid. The 'Solar-Plus' house has already been demonstrated, for example by the house of architect Rolf Disch in Freiburg. Even if solar thermal energy generation is not adopted, heating from a heat pump driven by electricity from photovoltaics is an alternative. Also miniature cogeneration plants driven by biomass allow 100% renewable energy in a house.

Solar Thermal Power Plants

Of the various basic types of solar thermal power plant,[106] power plants with mirror collectors have advanced furthest, and they have been used for years in California. The first 14 MW unit became operational in 1984, producing electricity at a cost of 28 cents/kWh. The latest plant of this type, which became operational in 1992, has an efficiency of 16% and produces electricity for 9 cents/kWh.[107] These plants are the international showcase facilities for solar power generation, although 25% of their operation is still with oil-fired back-up generators. Further expansion was halted when the operator, LUZ International, went into bankruptcy in 1992 because the power purchasing contracts were linked to falling oil and gas prices and because Federal tax breaks ended at the same time. Nevertheless, these plants provide convincing evidence for the prospects of immediate, widespread market introduction.

This type cannot be used everywhere, however, because they require direct solar radiation of more than 1800 kWh/m^2 annually, which means that only high-insolation areas south of the 40th parallel that are largely free of mist and smog are suitable. They can be deployed in many parts of the world: from California to southern Italy or southern Spain, from India to Brazil, from Australia to Mexico. There is no good reason to wait for their broad-based introduction until some private investor is found, but it is exactly this type of politically motivated waiting that is occurring nowadays, with the result that the producer of the collector mirrors, Germany's Flachglas-Solar (now Pilkington Solar International), largely stopped production in the year of the Rio Conference. If we compare the power generation data of these solar plants with those of oil-fired plants, we find that, with power plant capacities of 80 MW and an oil price of $36/barrel, both types produce electricity at approximately the same cost. If the oil price drops below this level, solar-generated electricity is more expensive, however. These solar farms represent just the beginning: 350 MW total capacity is a lot, given the early stage of introduction of solar energy technology in general, but it is minimal compared with the capacities of conventional power generation facilities. Adding several thousand megawatts of solar capacity promises to lower cost even more, and is likely to result in further technological improvements, which means that in the foreseeable future we could potentially achieve power prices of 6 cents/kWh – a price that is unlikely to be matched by any coal, oil or nuclear power plant. One megawatt-hour of solar thermal energy production means 2,000 tons less CO_2 per year; the plants operating in California are already saving about 700,000 tons every year.

Combined with political initiatives to support the introduction of solar thermal technologies, it seems entirely possible to achieve a hundred or more times the present capacity within 10 years, doubling that number every 10 years subsequently. This would mean that during the next few decades about one quarter of worldwide power generation would be solar-based – in addition to other solar technologies in the same and other regions. In the process, it would be possible to phase out the currently used oil- and gas-fired back-up turbine generators. One study has calculated that for the Mediterranean area alone – in the sunbelt from southern Italy and southern Spain to North Africa and the Middle East – 6% of the land area (8.9 million km², or 3.4 million square miles) of the states bordering on the Mediterranean Sea would be suitable for solar thermal power plants, 80% of them located in North Africa.[108] Italy and Spain would be able to meet their entire electricity requirements with solar thermal power plants on their own soil. Undersea power cables could transport electricity from northern Africa to Europe, and from southern Europe to central Europe. A Swiss utility is already planning the construction of a solar thermal power plant in southern Spain to meet Swiss demand.

Some 5,000 MW of new power generation capacity is being built annually in the entire Mediterranean area; Italy alone is planning to add 20,000 MW generated in new gas- and oil-fired power plants by 2005, and another 50,000 MW by 2025. If only one third of these new facilities were solar thermal units, more than 30,000 MW in solar thermal capacity would be available in the Mediterranean region by 2020, including 8,000 MW in Italy – a number that would be equivalent to 16 million tons less CO_2 generated in Italy, equal to almost 5% of that country's current CO_2 output. The introduction of small solar thermal plants also offers great opportunities – for individual operators as well as for villages. In addition, solar thermal facilities can achieve process temperatures of several thousand degrees Celsius, making them suitable for appropriate industrial pro-cesses such as iron production.

Photovoltaics

Some 120 MW of photovoltaic (PV) solar cells are being produced every year worldwide for the direct conversion of solar radiation into electric power; a grand total of about 700 MW have been produced at the time of writing. But so far there is no mass production from automated assembly lines. The investments are too large for small companies, and large companies have not launched projects of this type for a long time. However, in the meantime, some initiatives have begun in Japan and the United States and in Europe through subsidiaries of BP and Shell (BP-Solar and Shell-Solar). They are being forced into it by numerous initiatives from small and medium-sized companies, such as Solarfabrik Freiburg or Ersol, which have dared to attempt what larger companies would not undertake.

Nonetheless, the average sales price of PV cells dropped by two thirds between 1990 and 1997, while the conversion efficiency of current commercial cells increased from 4% (1978) to about 15%. There are already cells with more than 20% efficiency. In 1990, the price for PV power generation in Germany ranged from $0.87 to $1.10/kWh without support systems. In the meantime the price has fallen to just over $0.55/kWh. These calculations are based on a very low estimate of the operating lifetime of solar cells. Assuming realistic operating lifetimes of 30 years, prices of well under $0.50/kwh become possible, even in Central Europe. The conventional, frequently repeated arguments as to why these energy carriers can be deployed only conditionally, despite their encouraging perspectives, generally run like this: it would be possible to lower the cost of solar cells, but not of the technical support structures (racks); solar insolation would be too variable, requiring spare capacities by other energy carriers for the downtimes (with their costs needing to be added to the pure PV costs); PV would require too much space.

Numerous examples already exist that demonstrate successful integration of such cells into building components: solar cells are being installed not on facades or roofs, but rather as a solar cell skin instead of facades or roof shingles. This eliminates the costs of cell racks as well as costs for facades or roofs. The argument that support structures are too expensive fades into oblivion if existing buildings are used as solar fields. It is also economic nonsense to use real estate, which can be used for other purposes, if enough building surfaces are available for solar panels. This effectively collapses the argument of excessive land requirements.

Photovoltaic utilization is even more attractive under the so-called island concept: that is, by saving the costs of connecting to the grid. This approach is more economical for lighting traffic signs in remote rural areas, for example, or for electricity supply for villages in the Third World.[109]

How much PV potential could actually be available is documented by a study of the Newcastle Photovoltaic Application Centre.[110] Once photovoltaics are integrated into existing buildings, 10% of the existing stock of building surfaces would be sufficient to meet Great Britain's entire electricity needs. The study looked only at the realistically available total – in other words, the variations in daily and seasonal solar fluctuations on the one hand and daily demand fluctuations on the other. This 10% building surface figure is sufficient even in winter, concluded the study, with only winter evenings presenting difficulties. The average costs of expanding the entire system would be equal to the current average cost of electricity in Great Britain: about 8 pence/kWh. Once installed, these solar cell facilities could deliver 165,000 MW between April and September, only 35,000 MW between October and March, and, finally, only 27,000 MW in December. In other words, summer energy would have to be stored for the winter, which would increase costs. All in all, though, this system could offer 30% more electric power in the target year 2020 than Great Britain consumed in 1989.

The calculations of this British research group should not be mistaken for a practical blueprint for the introduction of solar cells, however. It is neither necessary nor sensible to place all bets on a single solar energy carrier. However, even if we discount some of these claims for Great Britain, one thing is certain: while such a forecast may be a bit too optimistic – an entire country supplied with electricity from PV alone by 2020 – it becomes realistic if we think in terms of adding other solar energy carriers to photovoltaics. And even if that should turn out to be impossible by 2020, it certainly should be achievable by 2030, 2040 or, at the latest, 2050. But most of all: what seems possible for a relatively sun-starved Great Britain should be possible in almost all other countries.

The fact that there is no automated assembly line production of solar cells so far, and that too little silicon material is produced specifically for solar cell manufacture (silicon currently used for PV production is a by-product of silicon production for computer chips), is illustrative of industry's restraint. Only a shift to modern volume production will signal any real start for the various types of solar cell under consideration, whether they be thick-film cells with high material requirements, or as a German scientist, Werner H. Bloss, emphasizes, thin-film cells with less efficiency but also much lower material requirements, which could take the place of, for example, the innumerable glass facades of large office buildings.[111] But the start of mass production requires investments of several hundred million dollars – too much for small manufacturers, and the large producers and their banks have not yet shown any interest.

The frequently cited argument that the market prospects are still too uncertain has been disproved by reality: demand has begun to outstrip production. If we compare the pitiful attempts – at both political and corporate level – to develop PV technology on an industrial scale with the efforts to develop ever more powerful computer systems, the difference in priorities becomes glaringly clear. Both solar cell modules and computer chips employ semiconductor technologies, but the first category is always disparaged with references to the allegedly excessive cost of electricity production. Nobody asks about the costs of developing new computer chips. Instead, the economy's future competitiveness is invoked, without any questions about the economic viability of deploying such high-powered computers, and, on the other hand, without taking notice of the future economic significance of photovoltaics.

Wind Power

Generating electricity from wind only became of interest again in the 1980s. Some 6,000 MW had been installed worldwide by 1996, representing three times the 1990 level. But while research efforts have been minimal so far, and production numbers have been small, the cost of wind-generated electricity at certain favourable sites is already comparable to the cost of nuclear power! Since these outlays consist mainly of equipment costs, overall expenditures can be reduced, above all, by increased production, technical advances and intelligent utilization of sites with the required wind speed.

In Europe, costs of less than 6 cents/kWh have been achieved with modern equipment and in good locations; on average, the costs are comparable to those of a coal-fired power plant equipped with flue gas elimination equipment.

With this level of development already achieved, there is no longer any reason to exercise restraint. The potential is large enough for much more than a mere 10% or 20% of the electricity supply. For western Germany alone, it has been calculated that there are some 20,000 km² (7,700 square miles) suitable for wind power. With full utilization, this would equal some 250,000 MW, or 1 million facilities, or twice the output of all German power plants.[112] More recent analyses suggest that the number of suitable sites is actually increasing rather than decreasing, and assuming further technological development, areas with lower average wind speeds become promising. Additionally, there is a considerable offshore potential, with wind farms located in waters near the shore. The actual land consumption of wind turbines is minimal, and amounts to less than 1% of the areas regarded as suitable – in other words, of these 20,000 km² (7,700 square miles), only about 200 km² (77 square miles) are actually needed. Agricultural operations – or fisheries – can continue in the shadow of such facilities. The Danish and German examples demonstrate how this theory can be put into practice. In Germany there was an expansion from less than 100 MW to 6,000 MW between 1991 and 2000, thanks mainly to the effect of market forces introduced by new legislation governing electricity supply, called the Renewable Energy Act. The Danish government plans to achieve a 35% share of the country's electricity provision coming from wind power by 2015.

No form of energy generation needs less acreage than wind energy, which makes the landscape conservation argument voiced by some opponents almost embarrassing. The bird protection argument that is occasionally trotted out is fallacious as well, for thousands of birds die on high-voltage lines or are driven away by filthy air or climate changes caused by conventional energy consumption. Around the end of the 19th century there were some 100,000 windmills operating on the North Sea coast between Holland and Denmark, 30,000 of them in Denmark's interior alone. To the extent that they have survived, windmills are regarded as an integral part of the natural landscape. Why could that not become the norm once more? The nature conservation and land preservation arguments against wind power facilities are far-fetched. Nevertheless, they are disseminated with extreme agitation, accompanied by the benevolent approval of the large energy producers, who, as the creators of smog, have invented the term 'aesthetic environmental pollution' or who talk about noise pollution, despite the fact that a large wind turbine creates less noise than the average truck. The Danish wind plants, which are the best example so far, repudiate all the hot air against wind power.

More importantly, given the constantly improving question of costs and the decentralization aspects that are part and parcel of this type of energy utilization, large opportunities for new economic initiatives are opening up for the operators. These operators would become independent of other energy producers, and in turn would help to elevate the general level of economic development outside conurbations. Wind power offers the opportunity of entering another field of economic activity to farmers in coastal areas and in mountainous regions, and could thus improve these regions' economic infrastructure. The political job would consist mainly in helping to remove narrow-minded administrative obstacles.

Renewable Biomass

The potential for biomass as a source of energy will be opened up only when the energy chain of nature's carbon cycle, from photosynthesis to oxidation, is once more linked intelligently and effectively to the supply of technically usable energy. The largest share of bio-energy, which is stored in agricultural and forestry products (that is, solar energy chemically stored in organic carbon compounds), is not used energetically, which means that its potential usefulness has turned into an additional environmental burden. In developing countries, biomass is used largely as fuel, often destroying existing resource stocks by failure to apply sustainable principles, with the final result of 'devegetation, deforestation and desertification'. Whether and how we use biomass is an indicator of whether and how man destroys nature or lives with it in harmony.

Biomass is an extensive solar energy carrier if its capacity for regeneration is utilized: provided that stocks are preserved by cultivation, biomass is available worldwide and without time limits. For this basic reason, it has the potential to make an enormous contribution to the replacement of fossil and nuclear energy sources. Depending on its origin and type of processing, biomass, continually replanted in natural material cycles, can deliver technically exploitable energy raw material in all forms – solid, liquid or gaseous:

- Solid biofuels (for example, wood and straw) are used in stoves, heating furnaces and cogeneration plants for the production of heat and electrical power, or are upgraded by pyrolysis to wood gas to power internal combustion engines.
- Liquid biofuels (for example, pure vegetable oil and alcohol) are used to fuel internal combustion engines for cars and trucks, or to generate heat and electrical power in generators or cogeneration plants.
- Biogas, a product of methane fermentation of green biomass or organic wastes, consisting largely of methane as energy carrier, is also used to generate thermal, mechanical and electrical energy.

One major advantage is that biomass can be used as fuel for transport systems with neutral environmental impact.

Its second major advantage is that it is the only solar energy carrier that is as easy to store as crude oil, natural gas or coal. Its third advantage is that, in addition to its climate-protecting aspects, it offers the largest number of additional beneficial effects: it promotes vegetation, instead of destroying it; it offers new income opportunities for agriculture; rural areas would experience new economic stimuli, which would help to offset the constant expansion of conurbation areas; organic waste can be removed and thus contribute to increased hygiene; and plants provide biological materials for the pharmaceutical and chemical industries as well as for the construction material and paper industries that can be recycled by nature.

Biomass as solar energy carrier consists of innumerable individual energy carriers, analogous to its manifold organic origins, which require different technologies for their processing and conversion. Generally, two categories can be distinguished:

- Organic residue materials and waste products, which occur in large quantities in our economic system and which release their energy as well as the metabolic products CO_2 and methane without any energetic utilization via slow, cold oxidation (rotting processes). This includes most of the materials used for biogas production (including liquid manure, foliage plants, pressed residues, and slaughterhouse scraps), organic domestic rubbish and sludge, as well as those organic wastes from agricultural and industrial production usable as solid fuels (such as straw, grain and rice husks, wood chips, construction rubble, old furniture, and waste paper).

- Energy plants that have been cultivated for the sole purpose of energetic utilization or as industrial raw materials – from the planting of fast-growing forests to palm oil plantations, sugar, millet, or to the cultivation of reeds.

There are examples that are both remarkable as well as controversial, such as the Brazilian alcohol fuel programme. At present, some 100 million litres (26.4 million US gallons) of ethanol are made from sugar cane, powering some 4 million cars. The advantages (CO_2 reduction, creation of 500,000 jobs, improved national balance of payments by reduction of energy imports) are in part offset by disadvantages such as the expulsion of small farmers, subsidizing of the build-up of the ethanol industry and, implicitly, of the owners of automobiles at the expense of society's poorest strata – disadvantages that are inextricably linked to the country's social structure, however.[113] Additionally, there are undisputed positive examples such as Sweden and Austria, which already generate well over 10% of their energy supply from biomass, mostly from agricultural wastes and forest products. Some combined heat and power production plants in Sweden, such as the *Brista-Kraft*, which was awarded the European Solar prize by EUROSOLAR in 1997, have 55 MW of thermal and 25 MW of electrical capacity.

While the utilization of biological wastes offers only advantages, using energy crops can create problems unless these are correctly applied or selected – rather as can happen with intensified food production via the excessive use of fertilizers and pesticides. Nevertheless, it is easier to avoid these mistakes at the start of such a project than try to correct them later on. There is no such thing as a lack of agricultural space, with adverse effects on food production. Farmland is constantly being taken out of production in the European

Table 12

	Present cropland	Potential cropland	Degraded land
Latin America	179.2	889.6	188
Africa	178.8	752.7	249
Asia (ex. China)	348.3	412.5	319
Total	706.3	2054.9	756

Community to reduce over-production, and some farms are closed down because their owners see no future in farming (in part because of the removal of national barriers against agricultural imports from Eastern Europe or Africa). The list in Table 12 gives an idea of the usable agricultural land in 91 selected developing countries (in millions of hectares).[114]

In theory it is possible to meet our entire energy demand from biomass, but this is unnecessary because of the existence of other solar energy carriers. Calculations of the total potential have indicated that a mere one third of all plant waste could meet 10% of the world energy demand.[115] One study by the Forward Studies Unit of the Commission of the European Communities indicated that an area of 40,000 km^2 (15,440 square miles) in Europe could meet up to 30% of Europe's energy needs.[116] Another calculation, by European Community expert Giuliano Grassi, arrives at figures of, respectively, 80,000 km^2 (30,900 square miles) and 70% of Europe's energy needs, based on the potential of fast-growing C$_3$ plants (eucalyptus produces 15 tons of dry mass on 1 hectare (2.47 acres)) and even faster-growing C$_4$ plants (for example, miscanthus with 35 tons/ha and sugar millet with 28 tons/ha dry mass, which can produce one third of that total as fuel).[117] Palm plantations laid out as energy farms and spread over 3 million km^2 (1.16 million square miles) could equal the annual production of crude oil. It has been calculated for Sweden that cultivation of willows could produce an annual dry volume of 6–8 tons/ha.[118] The most efficient utilization requires modern methods to extract oil from dry mass, and to ferment or pulverize the rest, turning it into solid fuels or biogas.

The widest range of applications and cost-effectiveness is supplied by C$_4$ plants, something that has led Franz Alt to coin the slogan 'Reeds instead of nuclear power'. Only 0.5% of the world's flora consists of plants suitable for food production. Instead of using these species as energy plants, it would be more sensible to selectively target forest plants suitable for energy use, based on the following criteria: fast growth, high energy yield, low water requirements, and low or no need for fertilizers. Grass and so-called C_4 plants are best suited for that. Most plants are C_3 plants, which cold-combust large parts of the carbon generated by photosynthesis and immediately release it again as CO_2 to the atmosphere, resulting in the loss of potential biomass. C_4 plants, on the other hand, can bind more carbon. According to studies by Wolfgang Ständer, there are more than 1,700 such species in eastern Asia, North America (notably Alaska), South America and Africa. Most of them can grow just as well in Europe, as evidenced by maize, which also belongs to this category. They possess the advantages of high yields, favourable cycling of nutrients into their root systems during the winter, low fertilizer requirements, and prevention of soil erosion. They offer the prospect of soil regeneration and a variety of growing methods, and thus help to avoid monocultures. Demonstrations have shown that they require only about half the amount of water compared with traditional agricultural and forest plants. They also offer a variety of uses: as energy; for chemical purposes, such as the manufacture of varnish and lacquer, solvents and synthetic materials; for the production of biological construction materials; and for the production of paper or packaging materials.[119]

Also significant is the use of biological residues from domestic households, agriculture and forestry operations. The production of paper, for example, could be based on the harvesting of natural forest waste. More than fifty developing countries would be able to produce as much energy from the residues generated by sugar production as they use now via imported oil.[120] Rice husks can be turned into briquettes for fuel.[121] Five tons of rice produce one ton of husks; if not utilized in some fashion and instead allowed to rot, they release methane (32 times more damaging to the atmosphere than CO_2). If used as energy, methane is rendered harmless in terms of climate. Whether coffee or peanut husks, coconut shells, olive or avocado pits, cotton stalks or potato green matter, almost all food vegetables grow a by-product that can be used for energy. Harvest by-products that can be utilized as energy amount to 4.5 tons per ton of coconut; 5 tons for cotton, 4.9 tons for millet and 1.8 tons for wheat.[122] This means that agricultural activities can be expanded from food production to energy raw material production including energy supply for the end-consumer. The energetic utilization of biomass thus provides an economic incentive to return agriculture to a biologically full-cycle economy.

The multiple use of the biomass economy is particularly important for the Third World. In agro-forestry, fuel production, food and feed production, and the production of industrially utilizable materials (for medicine or construction lumber) can be combined with soil conservation and regional climate protection. In the area of water plant cultivation – water hyacinths, for example – food, animal feed, recyclable fertilizers and energy for biogas facilities can be produced while cleaning up polluted bodies of water at the same time.

Biogas plants for the energetic utilization of liquid manure, bio wastes, sludge and domestic garbage show how double or triple problems can be solved with one operation. China boasts several million facilities that use biogas, mostly at low levels of technology. In Denmark, on the other hand, there are a number of more sophisticated processes for utilizing organic waste – as, for example, in Lintrop, where farmers jointly operate such a plant, which converts liquid manure, stomach and intestine content matter, fish waste and sludge into energy at competitive costs. The farmers save on their own energy costs, reduce oil consumption and sell electric power, in addition to generating revenue via fees for carting away other people's refuse. A statistical review of western Germany's agriculture indicated that 14 million working animals produce 146 million tons of energetically usable organic wastes annually, 40 million tons of sludge, and 5 million tons of organic wastes – a grand total of almost 200 million tons, equal in terms of energy potential to about 2% of the country's total energy requirements.[123] This amount may appear small, but it corresponds, after all, to about one fifth of the current share of nuclear energy.

Environmentalists are generally sceptical about the utilization of biomass – less about the use of biomass wastes than about the cultivation of energy plants. The critics argue that:

- even their combustion generates CO_2 emissions. This is valid only if there is no closed production and harvest cycle – 'put and take, take and put' – and when plants are harvested without replacement. Biomass, by returning to nature's cycle via either 'cold' or 'hot' rotting processes, resupplies its carbon to new plant growth in the form of CO_2. If this were not happening, the world's plant growth would very quickly come to a halt. In contrast to fossil carbonaceous energy carriers, biomass produces during its 'living phase' the oxygen required for its reduction, and leaves behind only that amount of CO_2 that it removed from the atmosphere earlier. This is the cycle that forms the basis for the concept of energy plants: in effect, the amount of CO_2 generated during combustion is immediately reabsorbed by the new plant growth, creating a natural CO_2 balance.

- soil erosion is being promoted by the expansion of the nature-damaging agricultural industry, with the massive use of the artificial fertilizers and pesticides that make large monocultures possible. This type of production is avoidable, but it can also lead to conflicts like the one between intensive and extensive cultivation in food production. The fact that current agricultural practices operate mainly in a nature-destroying manner does not halt food production either. Instead, a more environmentally friendly kind of agriculture must be introduced. The same holds true for biomass utilization for energy purposes. Negative examples – such as rape cultivation currrently promoted in the European Community and, in some instances, the production of alcohol as fuel – are rashly and simplistically applied to biomass as a whole. These examples represent projects where environmental criteria were not to the fore, and which were launched with insufficient advance planning. Table 13 on page 115 compares the environmental effects of the biofuel cycle with those of coal.

Because of the greater variety of suitable plants, there are many more possibilities for avoiding negative effects on the soil than with food production. If degraded land is cultivated, there is even the potential for upgrading soils through the cultivation of energy plants – by encouraging infiltration by water, by controlling erosion processes, and by reducing the danger of extensive fire. By mixing plants for the most varied purposes – food production, plants for energy production, plant-derived raw materials – there is even a unique opportunity to replace monocultures with multicultures and help to increase bio-diversity. High land area requirements are not a serious

Table 13

Environmental aspects of the biofuel cycle	Net effects of energy crops compared with coal for power				
	Good	Fair	Neutral	Poor	Bad
CO_2	X				
SO_2	X				
NO_x		X			
Heavy metals	X				
Pesticides		X			
CO			X		
CH_4		X			
Nitrates, nitrites, ammonia		X			
Phosphates		X			
Chlorine			X		
Alkali metals				X	
Dioxins			X		
Hydrocarbons			X		
Volatile organic compounds		X			
Aldehydes			X		
Iso- and monoterpenes				X	
Refuge value		X			
Local biodiversity and habitat	X				
Non-local biodiversity and habitat		X			
Game species productivity		X			
Water discharge		X			
Pest and disease dynamics			X		
Change in genetics of background plant population			X		
Soil organic matter		X			
Soil movement		X			
Soil productivity (many factors)		X			

Source: J.W. Ranney, *Principles and Issues of Biomass Crops and the Environment*. Oak Ridge National Laboratory.

argument: depletion of land should be understood to mean excessive construction on the land, not its biologically appropriate utilization.

Hydro Power

Less than 10% of the world's hydro power potential has been exploited so far. Some 650,000 MW – 18% of the world's total electricity production – are produced by large hydroelectric stations, plus another 25,000 MW by small hydro plants. Because of the large amount of water being built up and the land required, large facilities can accelerate soil erosion, modify regional climate, and have negative tectonic and seismic effects. Their construction frequently leads to the dislocation of many peoples, and the breaking or destruction of a dam can cause flood catastrophes. These large facilities, for which there is no more room in Europe, are still less problematic than nuclear or fossil power plants, but it is not necessary to continue building facilities of this type. They require long and expensive power transmission lines, especially in developing countries, turning them into expressions of a development model that centralizes social structures. Small hydro power plants have good prospects, however, as have so-called 'micro' hydro plants, which would benefit from further technical development.[124] In Germany alone there are many thousands of unused or disused water facilities that could be brought back into operation. Often the same unreasonable, imbalanced objections have been raised against their utilization, however, as against wind power facilities.

Tidal Energy

A 240 MW tide power plant, La Rance, exists in northern France. Canada is investigating prospects of multi-megawatt facilities in the Bay of Fundy. Great Britain discussed a power plant stretching 16 km (10 miles) that would generate 7,000 MW (equal to 5% of Britain's power supply), at costs of 4–6 cents/kWh. Large facilities can cause problems, however, by changing tide levels, the coastal geography and marine biology; one technical problem, for example, is silting. While these problems are not comparable to those of nuclear or fossil energy use, nevertheless they are significant enough to relegate large tidal power plants to a secondary position within the total renewable energy arena. If other potentials were fully exploited there would be little demand for tidal power.

Wave Energy

A circle of buoys located a few miles offshore could convert wave energy into electrical energy. Calculations indicate that Great Britain could potentially install 120,000 MW on its western coast alone, more than its total current power consumption. Costs of 9 cents/kWh have been calculated for an existing facility near Kvaerner-Brug in northern Norway. All facilities in operation thus far have capacities of less than 50 kW. None of the plants represents an environmental problem. Both their potential and their development possibilities have been largely underestimated up to now. They have the significant advantage that, although the buoys need to be tied down, they float on the water and do not need any bottom-anchored support structures.

Geothermal Energy and Ocean Heat

Geothermal reserves are limited and not necessarily renewable. If pressure and temperatures drop in a tapped reservoir after a period of heat withdrawal, it is exhausted for energy utilization. This form of energy utilization is also more environmentally friendly than fossil or nuclear energy, but its environmental quality is not comparable to that of solar energies. Geothermal steam may also contain toxic elements such as arsenic. Nevertheless, geothermal energy is an energy source with clear environmental advantages over fossil and nuclear energies, and it can play a significant supplemental role in certain regions. If we are talking about the actual process used for extracting geothermal energy, which is the heating of water through a system of water ducts in deep boreholes, then its environmental quality is very comparable to that of solar energy. The same applies to heat exchangers utilizing the earth's warmth on the earth's surface, which is a variant of solar thermal energy. The definition of environmental friendliness here is that the electricity for operating the host exchanger will come from renewable energies.

Ocean heat, on the other hand, is a variant of solar energy. Ocean thermal gradient power plants exploit the temperature difference between the ocean's surface and deep layers, and utilize the solar-heated surface layer of the oceans. The potential of ocean heat is more than 100 times the total energy requirements of mankind. Ocean heat could represent a solar 'reserve potential' but there are many reasons why its large-scale utilization may not become necessary.

Solar Hydrogen

A solar energy economy will have a larger share for electricity than the nuclear/fossil energy economy, but not all energetic processes that require fuels can be replaced by electricity and direct heat. A basic disadvantage of solar energy carriers is that – with the exception of biomass – they do not produce any fuels directly. While biomass energy carriers can substitute a large share of fossil energy resources because of their physical characteristics (solid, liquid, gaseous), solar hydrogen represents another option to guarantee the security of energy supplies.

There are three basic reasons why hydrogen must be produced electrolytically via solar-generated electricity:

- In instances in which even the combined supply of various solar energy carriers may contain the risk of downtimes, hydrogen can be utilized as 'stored sun' for additional power generation.
- The solar hydrogen component is required for economical use of solar-generated power: when we are approaching a level of power generation capacity sufficient to meet the demand of an entire country or large supply regions, there will be plants that at times – especially in the summer – generate considerably more electric power than can be distributed and than is immediately needed. In order not to waste this electricity, it makes sense to employ these capacities for hydrogen production and to use the hydrogen as energy, not for power supply purposes but as fuel, or for industrial process heat, in the steel industry, for example, and as a substitute for coal. The steel industry alone contributes about 10% to global CO_2 emissions. In Brazil, for example, its operations contribute to the destruction of the tropical rainforest because of its use of charcoal produced from wood from these endangered forests.
- The huge demand for hydrogen in the chemical industry has

been met so far with hydrogen produced from fossil raw materials and with fossil energy. Replacing this part of hydrogen production by water electrolysis driven by solar power – in other words, the splitting of water into hydrogen and oxygen – would represent nothing less than the 'greening' of the chemical industry. In the long term it would afford the opportunity to produce the petrochemicals needed by industry from the combination of solar hydrogen and oil derived from biomass. In the steel-producing sector, substituting hydrogen for coal would bring about an emission-free production process.

A number of arguments are being marshalled against the use of hydrogen:

- It is too dangerous because of its high explosivity. Its routine use over many decades in the chemical industry has proved that it is not an unusually dangerous material. Hydrogen gas is dangerous only in closed rooms; in open spaces it is probably less dangerous than petrol because it is lighter than air, and quickly escapes into the atmosphere.

- Its introduction would fail because of the lack of any infrastructure – for example, no refuelling stations yet exist. It is ludicrous to think that an infrastructure would be built for anything before there is a concrete demand for it.

- The introduction of hydrogen as fuel would open the door for hydrogen production from nuclear power. The arguments here are the same as those between nuclear and solar electricity.

- It is not mature enough for use in internal combustion engines because the fuel tanks needed are still too large. While this is still true for passenger cars, it is no reason for delay with trucks or city buses.

- Hydrogen would also produce nitric oxide emissions, because during the combustion of hydrogen with the air's oxygen not only is steam produced but also nitric oxides from the combination of atmospheric oxygen and atmospheric nitrogen. Certain catalysts permit hydrogen combustion at temperatures below 700 °C, temperatures at which the formation of nitric oxides is suppressed. This argument also becomes irrelevant with the closed-cycle combustion of hydrogen. Water electrolysis produces hydrogen and pure oxygen, which can also be stored and which can be combined again with hydrogen during combustion. Especially attractive would be the use of hydrogen with fuel cells (instead of standard motors) in large vehicles because their range would increase owing to the fuel cells' almost three-fold efficiency compared with that of conventional internal combustion engines.

But, most of all, detractors point to the high costs of hydrogen on top of the costs for solar energy. They also point to the dangers that they say would arise from centralization if hydrogen production facilities were sited in a few desert locations only – an argument that was at the centre of attention at the beginning of the hydrogen debate. However, there will be no need for central sites because, on the one hand, there is the prospect of an environmentally benign replacement of fuels by biomass, which will make a faster breakthrough because of cost advantages. On the other hand, production of solar hydrogen will have to take its cue from the efficiency criteria of, by then largely decentralized, solar electricity production. This means that economical hydrogen production probably does not start with some large solar plants somewhere in the desert, but more likely with smaller units in smaller areas with real cost advantages over large systems – with the help of electricity from hydro power, wind power or biogas plants, perhaps even in homes with decentralized electrolysis plants. Large facilities may not be added until later.

The introduction of solar hydrogen is faced with the particular problem that the technologies for both hydrogen conversion and hydrogen utilization still need to mature to the volume production stage. The dependence of one technology on the other further delays their introduction. For that reason, complete demonstration programmes would be helpful, beginning by using the cheapest currently available electricity from renewable sources: hydrogen production with electricity from hydro power plants, and utilization of this hydrogen in city buses, for instance. Where there is a solar energy supply system, producing seasonally more electricity than can be consumed, the surplus power could also be used to make hydrogen. Generally, as soon as grid-connected solar facilities begin generating more power seasonally than can be absorbed by the grid, the excess electricity should be used to produce hydrogen. If wind parks were combined directly with regional electrolysis plants, for example, the operators would be able to compensate for seasonal fluctuations in their energy supply with stored energy from their own hydrogen-powered reserve capacities. The extent of the demand for hydrogen would depend mostly on how much fuel could be supplied from biomass with sustainable methods.

The Solar Energy Mix

To assure the security of energy supply requires an 'energy mix'. You rely on several energy sources in order to avoid dependence on a single one. Precisely that goal is ideally achievable with solar energy. The parallel introduction of solar thermal and photovoltaic technologies, of wind power and biomass, of hydro power and solar hydrogen, ensures their mutual support in evening out fluctuations. This also includes the mixing of domestic solar energy carriers with imported solar energy. Solar energy's spectrum is much wider and technologically much more diverse than that of the conventional energy mix.

It is clear that the various solar energy carriers discussed here represent, in their totality, a potential that does not require each individual solar energy source to be exploited to the full in order to achieve the goal of a pure global solar energy economy. Even if one lowers expectations for each solar variant, the overall potential remains large enough to reach that goal and to satisfy all energy requirements in a few decades: see Table 14 on pages 122 and 123. The spectrum of renewable energy sources is considerably broader than that of nuclear and fossil energies. Their applications are more flexible and constantly adapted to geographical and bio-climatic conditions.

The cost calculations cited here for the various solar energy carriers do not include the storage costs that would occur with the use of wind power, photovoltaics and, in part, with the utilization of running water. Because of the discontinuity of this type of energy supply – so goes the constantly repeated counter-argument – reserve capacities of conventional energy carriers would have to be kept ready in order to meet continuous demand at times of low insolation, at night, and at times of low wind. The alternative, according to this reasoning, would be high storage costs in the form of solar hydrogen or large battery capacities. In any case, the opponents claim, all this would significantly increase the cost of solar electricity.

Table 14 *Characteristics and potential utilization of various*

Source	Sunlight	Direct solar heat I	Direct solar heat II	Environmental heat	Wind
Natural potential	Extremely high	Geographically limited	Very high	Very high	Very high
Geographical availability	Everywhere	Arid and semi-arid zones	Everywhere	Everywhere	Flat terrain, coastal and mountain areas
Continuity	Daylight hours, weather-dependent, seasonal	Daylight hours, partially weather-dependent	Daylight hours, weather-dependent, seasonal	Constant	Variable
Storage	Grid, battery, hydrogen	Grid, hydrogen	Water storage	Not required	Grid, battery, hydrogen
Utilization techniques	Photovoltaics	Solar-thermal power stations	Solar collectors, 'passive' solar utilization in buildings, pumps	Heat exchangers with electricity	Wind farms wind pumps
Central/ decentralized	Decentralized	Centralized	Decentralized	Decentralized	Decentralized, moderately centralized
Emissions	None	None	None	None	None
Land requirement	Possible, but not essential, (integrated with buildings)	Necessary, but using otherwise unusable areas	Possible, but not essential (integrated with buildings)	Not required	Slight; surface below rotors can be used for other purposes

renewable energy sources on the five continents

Biomass	Small hydro	Large hydro	Wave power	Tidal power	Geo-thermal
Very high	Limited	Limited	High	High	High
All non-arid regions	All non-arid and mountainous regions	Mountainous regions	Coastal waters	Coastal waters of oceans	Dependent on earth tectonics
Constant	Mostly constant	Mostly constant	Variable	Predictably variable	Constant
As for fossil energy	Grid	Grid, reservoirs	Grid	Grid	Not required
As for fossil energy	Turbines	Turbines	Wave power facilities	Tidal power facilities	Steam utilization techniques, thermodynamic water circuits
Decentralized, moderately centralized	Decentralized, moderately centralized	Moderately centralized, centralized	Decentralized	Centralized	Moderately centralized, centralized
CO_2-neutral, otherwise less than for fossil energy	None	None	None	None	Salts and dissolved gases
Agricultural and forest land, extensive cultivation supported	Slight	Large	None	Influence on local aquatic life	Slight

If we examine this argument with any sense of imagination, we find that it can be largely refuted. During the period (which can be expected to be quite lengthy) in which solar energy was phased in to present energy consumption patterns, existing conventional fossil-based power plants would still be available to compensate for any shortages. During this period, the power grid itself, into which solar power would be fed, would perform the role of a storage system: therefore no additional investment costs would be required beyond the immediate outlay for the solar power plants themselves.

If the attempt were made to phase in not only one element of solar energy but several at roughly the same time – wind power, photovoltaics and biomass, for example – the storage problem would become even less problematical; the various solar energy potentials would balance each other out. Biomass-generated electricity would be available continuously, in any case. If necessary, solar thermal-generated electricity could be used for meeting base loads, via imports by countries unsuitable for solar generation.[125] Any remaining downtimes could be bridged with biomass potentials and with existing pumped storage plants – and only after that, with solar hydrogen. Solar thermal heat is, in any event, storable in long-term storage systems for winter periods. In other words, if one were to rely solely on photovoltaics or wind power in a supply system, storage costs would be high. But the more diverse the solar energy mix, the smaller the storage problem – and hydrogen may not be needed at all for power generation, but merely for fuel requirements in heat engines, internal combustion engines, gas turbines, jet engines, and in industry.

Along the way to a solar energy mix we could strive for a combination of power and heat production via cogeneration plants with conventional energy carriers and solar energy – and we could increasingly switch to Stirling engines. These transfer external heat to a working gas inside the engine, which drives the pistons. This movement in turn generates electricity. Since a large amount of waste heat is generated with conventional energy carriers, this could be used for decentralized power generation. At a later stage, the energy needed for these Stirling engines can be produced solar thermally.

The most important characteristics of any energy carrier have been defined by Othmar Heise as being that they should be capable of meeting demand, user-friendly, clean, renewable and

Table 15

	Nuclear/fossil energy mix	Solar energy mix
Meeting demand	Currently yes, in future no because of resource scarcity	At present no, but yes in future because wealth of resources
User-friendly	Yes, but no with centralized supply structures	Yes, but with decentralized supply structures
Clean, without residue	No	Yes
Renewable	No	Yes
Accident-proof	Coal and gas, yes; oil (tanker accidents/spills, well fires) and nuclear power, no	Yes

accident-proof.[126] If we compare the conventional nuclear/fossil energy mix with the desired solar energy mix based on these criteria, it is obvious which is preferable (Table 15).

Solar energy carriers have none of the disadvantages inherent in conventional energy carriers, but will be able to match all their positive requirements. However, if 'appropriate use' is supposed to mean a centralized supply structure, solar energy would not be able to meet this requirement.

A number of starting points exist, complete with clear cost calculations. In other words, there is no reason to wait for yet more research and development. Even today, solar energy potentials exist that can compete with conventional energy carriers, even when only investment and operating costs are taken into consideration. The only relevant question is: does solar energy provide immediate advantages compared with other energy carriers if we introduce it today? Since the answer is an unequivocal 'yes', any further delay would be inexcusably remiss.

The Unexploited Potentials

Just as exciting as the quest for new energy sources is that for new utilization technologies. Energy sources require appropriate utilization technologies. Some traditional utilization technologies will become futile with the introduction of solar energy into civilian energy use, and new ones will become possible and necessary. Lines of development that have appeared in the history of technology, and which were neglected or forgotten, will be revived. The airship, for example – admittedly a slower, but environmentally more benign, alternative to the aeroplane – may be resurrected for cargo or holiday transport. Domestic appliance technology will undergo a basic transformation, as will large segments of industrial technology. The electric vehicle, ignored for decades, will replace present individual transportation modes based on the internal combustion engine, and it will derive its electricity from solar cells or wind power plants.

The unfolding of the solar age must occur with the same courageous drive as the 19th century introduction of the railways.[127] In the course of the general introduction of solar energy, the various components of solar energy technology will undergo changes as well. Like any other technology, solar technologies are capable of considerable development, enlarging the exploitable solar energy potential and making it more cost efficient. Key requirements for their development[128] include, among others:

- for solar thermal power plants, lower weight for support structures, more efficient mirrors, new mirror materials or improvements of radiation receivers, direct steam generation, use of storage units, and adaptation of Stirling motors to solar operations;

- for photovoltaic power generation, the development of material-, water- and energy-saving fully automated manufacturing processes, and of new materials for PV cells and for concentrators;

- for wind power, new and continued development of rotor blades, materials and aerodynamics for silent turbines and so on;

- for biomass, more experience to identify the most environmentally acceptable energy plants and to develop alternative methods for combustion technology, gasification and oil extraction.

One task that has been widely underrated is advancing the development of battery technology. All this is part of applied research that should not be left to public research institutes alone, but should be increasingly the province of commercial corporations. The possibilities for publicly supported research institutes in the development of these technologies are limited because they do not have any experience with volume production and the marketing of new products. Additionally, they do not have at their disposal the development funds that a large manufacturer with high turnover can marshal. The more successful a product is, the greater is the outlay for further improvements of that product. This is the main reason why large development efforts and quantum leaps of technologies typically occur only after their successful introduction to the market. If solar energies are already beginning to display remarkable efficiencies in some niche applications – even when measured with conventional economic yardsticks and despite putting solar research and development in second place – the collective experience of technology history gives an indication of the future potential of solar technologies.

The solar energy potential for industrial processes is largely unexplored, especially in the high-temperature area: for example, distillation, for surface treatment, alloys and process steam – with application opportunities especially in southern countries, in brickworks, in the glass and textile industries, in roasting houses, in the beverage industry and in drug manufacturing; or for heat generation with known chemical processes that could help to create the new scientific discipline of solar chemistry (for waste conversion, the elimination of toxic substances and manufacture of synthetic fuels, for instance). From lighting to engineering cooling technology, from light-powered machines to medical technology, vast new perspectives are opening up.[129]

These perspectives will become even more exciting once we initiate a wide spectrum of basic research for completely new opportunities for solar energy conversion. The ground still to be covered by basic research is immense and of the utmost importance for humanity – but still no broad-based basic solar research is carried out! Generally speaking, the next stage of basic solar research should follow the maxim 'Let's imitate the plants',[130] for photosynthetic energy conversion will provide us with vital new technologies, unknown today, for utilization of solar energy. Nature has developed many complicated catalytic systems that already function at ambient temperatures – a model for efficient electrolysis systems and for fuel cells. For example, electric fish are equipped with recyclable battery systems with high energy density, and could provide a model for the investigation of new types of battery. The field of botany provides examples for energy-saving methods, materials, insulation materials, syntheses of light-converting membranes, structures and energy conversion, that can be analysed and possibly reproduced technologically.[131] Future generations will no doubt smile at the unimaginative narrow-mindedness with which the elites of the 20th century approached solar energy – provided we manage to leap the many hurdles that stand in the way of solar energy . . . and provided that future generations are in a frame of mind to laugh about the past.

5

Solar Energy's Economic and Social Benefits

There are numerous indications that solar energy is far more than a mere stopgap measure to escape from the present environmental crisis. These include the natural as well as the developed – and still developing – technological potential of solar energy; the vast opportunities offered by abandoning destructive energy sources; and, not least, the new industrial perspectives arising from the conversion of our energy system. In addition to the environmental benefits, solar energy will bring about major economic and social gains. The creation of a solar energy system offers an unexpected and unique chance to release industrial society from the harmful consequences of the Industrial Revolution and to make available its positive accomplishments – particularly the social, democratic and cultural opportunities made possible by freeing mankind from slave labour – to all of mankind.

Destruction of the environment is the greatest danger for industrialized societies pursuing economic growth, but it is not the only one. The Western high culture of welfare states is evidently a thing of the past. Created by the pressure of social movements that emerged in the Industrial Revolution, they stabilized capitalism by making it more responsive to the social needs in its strongholds. But both old and new contradictions, as well as the growth of welfare costs, lead to the conclusion that the future of the industrial system is increasingly seen only in terms of jettisoning its social obligations. Political democracy will then once more be in danger. Modern history is unable to provide an example of a stable democracy based on permanent mass misery.

Economic and Natural Entropy

The validity of the law of entropy, not only for the use of traditional energy but also for economies in general, has been shown most clearly in the discussion by the economist Nicolas

Georgescu-Roegen. He identified the pervasive consequences of the Second Law of Thermodynamics if we continue to convert valuable finite materials into waste. These ideas were also hinted at in Wilhelm Ostwald's early energy-sociological analyses, which contained the seeds of these considerations, although he was accused at the time – in the early 20th century – of thinking solely in terms of energy conversion and ignoring materials conversion. Arguing against Ostwald's theses, the sociologist Max Weber, for example, maintained that

> for the production, distribution and utilization of the most important usable energies, the essential chemical and formation energy of those materials are just as irretrievably dispersed as is the case with all free energies according to the entropy laws – but, in contrast to the others, in historically measurable time periods.[132]

Energy and natural scientists as well as economists have gradually lost sight of these relationships, the recognition of which was once automatic for those claiming a deep scientific understanding. Georgescu-Roegen elaborated that the continuously growing conversion of energy and materials did not inevitably bring about constant further increases in productivity and, with it, continued growth, but that it would lead in fact to an existential crisis. The more that limited supplies are exhausted, the more expensive their exploitation becomes, and the larger subsequent problems loom, growing to tidal wave proportions. For a while, the developed industrial societies were able to let the tide of adverse effects move outwards, inundating others – mostly via the globalization of economic processes – but the flood tide is returning. The increasing costs of economic reproduction and the conditions of this reproduction – as the economist Elmar Altvater put it, the 'non-productive side of productivity increases'[133] – lead to a permanent economic and political crisis: to the 'Entropy State', in Hazel Henderson's appropriate phrase.[134]

Indeed, governments are busily working round the clock to mitigate the pervasive consequences of the thoughtlessly one-sided priorities of past decades, making it impossible for them to pay attention to, and spend money on, the much more important tasks of the future. Inevitably, this leads to further expansion of public expenditure and public works, a growing government bureaucracy, and an increasing number of regulations for business. This stimulates the demand for private and public services, leading to a further decrease in the proportion of individuals who actually produce something. As industrial production was made more efficient – not matched by growth in industrial production as such – jobs that were lost were, for a time, replaced by jobs in the service sector. But it is impossible to extend this trend ad infinitum because services have to be financed by industrial output. Financing is becoming more difficult, in part because the productivity of personnel-based services cannot be increased to the same extent as in industrial manufacturing activities. Largely for this reason, the service sector becomes constantly more expensive.[135] Once industrial growth begins to slow down as well, services become unaffordable and must be reduced: the permanent crisis has arrived.

In the decades of rapid growth and energy consumption after World War II, larger, more monopolistic and increasingly globalized corporate structures than ever before began to take shape. The effect was that national economies became inflexible and incapable of reacting to brand new challenges, because entire sectors of the economy that had grown too large would be faced with unacceptably large losses. The overall effect is that the contribution to economic modernization made initially by these corporations with improved efficiency is being reversed, leading to the conservation of outdated structures. Nevertheless, these corporations have enough influence to keep their heads above water with subsidies and nonsensical government contracts, because they can threaten job losses. Because of the increasing national debt, the flagship companies of former public industrial projects in military and aerospace technologies are becoming increasingly shaky – and that in a period of market saturation in industrial countries by products that used to represent the foundation of the boom: cars, TV sets, video recorders, refrigerators and so on.[136] But the answers presented so far are not aimed at new government or production goals, but merely at economic optimization of existing structures – by opening up new markets for the same products, by diversification of the corporate product palette, and by other means of increasing economic efficiency, as recommended by a growing sector of corporate consultancies and business schools. Yet even on international markets that are supposed to be tapped, purchasing power is declining and a lack of hard currency thwarts these efforts. It comes down to an unresolvable contradiction: initially, dependent and less developed economies are rapaciously exploited, and then these same economies are supposed to serve as new markets for the expansion of existing industrial capacities.

Increases in productivity without new production goals merely mean a shift of the problems of the corporations to the general public. Political obstacles to growth – social legislation, corporate taxes, environmental regulations – are supposed to be eliminated again; the influence of unions is supposed to be reduced; and energy is supposed to remain cheap in order to cement, via an 'escape into the future', the old success strategies. Practically no one among these would-be reformers is thinking about the contradiction between meeting the growing social costs implied by their recommendations and the simultaneous decline in government revenues. These social costs reflect these trends, and their growth makes evident the increasing gravity of the crisis. Only adequate, realistic calculation and analysis of these costs can lead to real economic rationality and, with it, to the possibility of effectively shaping the future. The simple-minded economic fixation purely on running costs and overheads has not only distracted from the social costs, but has also clouded the view of future social gains and the entirely predictable future economic gains from solar energy.

In terms of a serious debate on energy policy, the primitive economic comparisons between traditional energy and solar energy have been largely laid to rest. But this is not at all true in questions of energy policy and economics as practised today, where, as always, only the operating costs are taken into account. Notoriously overlooked is how many subsidies for conventional energy carriers were and are being consumed, and the degree of damage caused by the consumption of these energies. These social costs of the conventional energy supply system have now been recognized and, in part, scientifically calculated. The economist Hohmeyer calculated the social costs for power generation (taking into account environmental damage, an 'exploitation' surtax, publicly financed goods and services, subsidies and publicly financed research and development) as ranging from 1.944 to 9.975 cents/kWh for fossil fuels and from 6.288 to 43.83 as the amount that should be added to the price per total for electric energy in 1992. On the reverse side, he calculated a social net benefit (taking into account avoided social costs of conventional power generation, environmental damage, macroeconomic side effects and public payments) for photovoltaic power generation of 12.25–13 cents/kWh, which should be subtracted from the cost price. For wind power he calculated a social net benefit of 9.75–10.5 cents/kWh.[137] In practice, however, these links to consequences that far exceed the subject of energy supply are not taken into account. There is a continued blindness to the economic and social opportunities of solar energy, and a noisy chorus of so-called 'economic experts' who continue to oppose it.

Comparing Economic and Social Costs

There is embarrassing ignorance about solar energy's economic and social potential, even among some of those scientists who are the harshest critics of the traditional 'blindfolded' school of economics. Among them is Georgescu-Roegen, the modern father of an environmentally based theory of economics. As late as 1986 he wrote in a retrospective that the 'Promethean technology' he calls for to avoid economic and social entropy just does not exist yet:

> The direct use of solar energy does not fulfil the minimal strictly necessary condition of a Promethean recipe which is that some solar collectors could be reproduced only with the air of the energy they can harness . . . The main obstacle is the extremely weak radiation of the solar energy reaching the soil. The only upshot is that we need a disproportionate amount of matter to harness solar energy in some appreciable amount.[138]

This thesis is one of the broadsides that the proponents of the established energy system keep firing at solar energy. It is untenable, and was repudiated long ago. Even today, a solar cell has a useful life of 30 years or more; it collects in two years or less as much solar energy as is needed for all its production stages;[139] for wind power plants, it is half a year.[140] The material demand is no problem either: silicon, made from sand, is as plentiful as sand on the beach. Where solar facades, which are theoretically sufficient to meet the entire world demand for electric power, replace traditional building facades the amount of material required in addition to silicon production is no larger than for normal home or office building construction.

This type of thinking can and must be expanded, because it involves the 'third dimension' of economics beyond the production and distribution of goods: the transformation of energy and matter.[141] Solar energy will help to replace not only the entropy-accelerating conventional energy system, but also large shares of traditional mineral and chemical materials. These can be replaced by the agricultural production of raw and basic materials for synthetic processes – in other words, with biological materials, which are not described in this book. The recyclability of materials used is the counterpart to energy savings, which can be only a temporary solution to or easing of the problem. Only the replacement of these materials with biological ones can liberate us from entropic dangers if agricultural production follows the regeneration principle – in other words does not intensify use of the productive land areas, a practice that destroys soils in the medium term. Solar energy, which flows through the ecosphere and thus through matter as well, is therefore the source not only for a new energy system, but also for a new system of cycling matter.

There are other arguments that – consciously or unconsciously – reveal ignorance about solar energy's actual prospects, and which thereby distort the perspective. It is said, for example, that solar energy requires too much land. But we have seen in the previous chapter that this is not true: only by excluding the consumption of land required for conventional energy production in the overall equation can one possibly arrive at the blinkered conclusion that solar energy requires more acreage. It is also conveniently forgotten that the use of nuclear energy has already contaminated more land than will ever be required for solar cells and solar collectors. Solar energy's variability is criticized without taking into consideration the previously described solar energy mix. 'Energy density' is said to be too low – another erroneous argument, as will be shown later. Furthermore solar energy is said to produce 'environmental damage', an argument that conveniently overlooks a fundamental difference: the consumption of nuclear and fossil energies produces irreversible consequences for the entire ecosphere, but the use of solar technology can, at worst, produce a locally limited, reversible impact on the environment. Some solar energy opponents present themselves as the new breed of 'eco'-fundamentalists who, by defending an energy system of global destruction, are splitting hairs in their search for arguments against solar energy.

The high environmental costs of traditional energy carriers compared with solar energy, as cited by Hohmeyer – social costs of 4.96 cents/kWh for coal-derived electricity and up to 0.13 cents/kWh for nuclear electricity, compared with 0.0006 cents/kWh for wind power and 0.28 cents/kWh for photovoltaics – show clearly that the reference to environmental damage caused by solar energy is totally untenable. So far, the political institutions have been unwilling to take any meaningful steps, either nationally or internationally, to calculate the social costs of traditional energy. Their excuse was that this would entail national economic risks with incalculable consequences if energy prices were to increase. Evidently they are willing to live, now as before, with social damage as an unavoidable evil.

It is conceivable that additional aspects of solar energy may provide the overdue impetus for a solar strategy. The real social costs of conventional energy far exceed the figures that have been calculated so far, if one looks at the entire energy system with its far-reaching consequences. The social gains of solar energy are correspondingly large.

New Jobs

It has already been emphasized that the introduction of solar energy will generate new jobs in nearly all sectors of industrial production and the service sector. The objection is frequently heard that solar energy technologies would be too expensive for economies that are competing with each other, but this merely reveals an insular way of looking at economic issues. The cost of solar energy is higher not because of the cost of the primary energy but because of labour costs. But higher labour costs then mean: solar energy utilization requires more jobs! The introduction of solar energy would lead to a reduction of unemployment and, with it, the social costs of unemployment. The World Watch Institute has calculated that five times as many people would be employed for the production of 1,000 GWh from wind energy than from nuclear power (Table 16).

The numbers of jobs required would undoubtedly not be as large as they are today once solar energy plants were produced in volume. But their introduction would mobilize new development and production activities in many segments of industry, not only for the production of solar energy converters, but also of equipment for solar energy utilization. Crafts and small business would be given new impetus.

Table 16

Technology	Jobs (per 1000 GWh/year)
Nuclear electricity	100
Geothermally sourced electricity	112
Coal-derived electricity, including mining	116
Solar thermal electricity	248
Wind-generated electricity	542

Source: World Watch Institute

Reduction of Administrative Costs

As mentioned, administrative intervention in business, and with it the costs of traditional energy supply, increase because of growing environmental damage and other dangers. This extends into the international arena: for example, agreements about the protection of the oceans from waste products of energy generation or control of the nuclear fuel cycle. Hand in hand with these rules and regulations comes an administrative apparatus to control damage and monitor execution. The growing inadequacy in actually enforcing these rules points to increased administrative costs, which never show up in any energy calculation.

The introduction of solar energy offers the opportunity to reduce these administrative costs. This has not happened so far, partly because of real administrative obstacles to solar energy such as construction codes, land-use rules, and nature conservancy codes that, in misinterpreting the hierarchy of environmental dangers, occasionally raise more objections to the installation of solar energy facilities than to traditional energy equipment. But there is no need for special safety rules for solar energy plants, no emission limits, no waste disposal rules, smog regulations or monitoring stations. Only the utilization of biomass requires a few administrative regulations, which do not have to be any more extensive than those that exist already for agriculture and which can be handled by existing agricultural administration agencies.

Overall, solar energy represents a great opportunity for de-bureaucratization. The more solar energies are introduced, the lower the administrative effort required both by the government and by corporations. Solar energy therefore becomes a means for rationalizing and streamlining public expenditure.

Hard Currency Earnings and Reduction of Subsidies

Traditional energies cause high purchasing costs among energy-importing countries, and are a burden on the balance of payments.[142] The mobilization of solar energy permits a drastic increase in the available domestic energy sources in any country. This increases energy security and decreases international dependency. It also offers the opportunity to reduce agricultural subsidies drastically, as will be shown in Chapter 7.

Reduction of Military Costs

The need for the large Western energy-consuming countries to secure access to cheap sources of oil in the Near and Middle East is the main reason why the United States alone spends more than $60 billion per year on military tasks in the region.[143] The 1990/91 Gulf conflict and consequent war alone cost some $70 billion – a figure that does not appear on any oil bill. This sum is probably sufficient to create the basis for independence from oil imports by replacing them with plant-derived biological fuels. But even after the Gulf conflict, calculations were made in the United States that paying for the military security of the oil wells – based on the amount of oil imported from the Middle East – was $100 per barrel, in effect doubling the market price to the USA in 1999.[144] At present, these costs are still rising for members of NATO: the bloc's new strategies of forming rapid deployment forces that, even after the end of the East–West conflict, stand in the way of significant reductions in arms budgets, are justified as necessary to cope with new instabilities and dangers in the oil-producing countries of the Mediterranean region and the Middle East. Mobilizing solar energy would reduce – in fact, eliminate – the official premises for these outlays.

Additionally, the costs of fossil and nuclear energy continue to increase because of the growing dangers of the proliferation of nuclear weapons through civilian nuclear technology. Also, the continued stabilization of feudal government structures in oil-producing countries – the quid pro quo for generous energy supplies at favourable prices – and the resulting social tensions are part of the cost calculations of the dominant energy system, together with social catastrophes in developing countries because of the lack of energy and the resulting streams of refugees. A clear, massive change in priorities favouring solar energies would go a long way towards alleviating these problems.

Increase in Land Use

It has already been shown that economically appropriate use of solar energies reduces the amount of land needed for energy supply. Once solar cells on roofs and wind farms on grazing land replace traditional power plants, including the mining regions for primary energy, additional land will be gained instead of lost!

Solar energy use will eliminate the contradiction between economic growth and destruction of nature so that a new economic dynamic of its own will be created. There is no market potentially larger than that for solar technologies, and so far it has been almost completely neglected. Entire new segments of industry can be created, segments that would not face any fundamental acceptance problems by society, as is increasingly the case with traditional centres of production. This, in turn, guarantees security for new investments.

The economic and structural change triggered by the introduction of solar energy will be characterized by a drastic increase in the number of wealth-creating jobs. Traditional jobs in energy production and energy supply will be reduced, including transport and distribution, and these will be replaced by productive work in manufacturing solar energy panels and the installation trades.

Linked to this is the opportunity to reverse existing trends: a move away from service jobs to productive work. This trend will be augmented by the fact that, with a growing number of people generating their own energy, another element of the existing non-productive service jobs will be reduced. Similar considerations apply to health services, since undoubtedly solar energy will contribute to an improvement in the general level of health. One US cost estimate has found that the conventional energy system causes corrosion damage of about $2 billion annually, health costs of $12–82 billion, agricultural damage of $2.5–7.5 billion, military security costs of $15–60 billion, unemployment costs of more than $30 billion, and subsidies of various types of $55 billion – in other words, hidden costs of between $116 billion and $236 billion, depending on how rigorous the calculations are.[145]

The dominant energy system creates a number of other additional future costs as well as long-term environmental damage. This is true especially for the combustion, and therefore destruction, of natural resources that have substantial importance for non-energy purposes, such as industrial raw materials and pharmaceuticals. Solar energy is not the only factor, but probably the most significant one, aiding resource conservation that will prevent future disruptions in many industry segments. Solar energies offer the opportunity for an economic and environmental humanization of the Industrial Revolution and for administrative and social reforms that can no longer be realized under current conditions. Table 17 on page 140 contrasts the social cost factors of conventional energies with the social gains of solar energy. It displays a surprising picture of the social benefits afforded by solar energy.

Solar Energy's Economic Advantages

Every technology has its own economic rules. For this reason, all those economic analyses that merely compare the first price of the various energy carriers without analysing the mix of their respective cost factors and their prospects for economic optimization are not particularly relevant. A more detailed look

Table 17 *A comparison of social costs and benefits of various energy carriers*

	Fossil energy	Nu-clear fission	Nu-clear fusion	Solar thermal heat	Solar thermal elec-tricity	PV	Wind	Biomass waste crops	Renewable biomass plants	Small hydro plants	Large hydro plants	Tides
Unlimited availability	No	No	Yes	Yes	Yes	Yes	Yes	Yes	Yes	Yes	Yes	Yes
Reduce CO₂	No	Yes	Yes	Yes	Yes	Yes	Yes	Yes	Yes	Yes	Yes	Yes
Reduce heat emissions	No	No	No	Yes	Yes	Yes	Yes	Yes	Yes	Yes	Yes	Yes
Land conservation	No	No	No	Yes	Yes	Yes	Yes	Yes	Yes	Yes	No	Yes, if correctly applied
Avoid danger of major accident	No	No	Maybe	Yes	Yes	Yes	Yes	Yes	Yes	Yes	No	Yes
Reduce administrive costs	No	No	No	Yes	Yes	Yes	Yes	Yes	No	Yes	No	No
Relieve balance of payments	No	Yes	Yes	Yes	Yes	Yes	Yes	Yes	Yes	Yes	Yes	No
Reduce international conflicts	No	No	No	Yes	Yes	Yes	Yes	Yes	Yes	Yes	Not always	Yes
Social acceptance	No	No	No	Yes	Yes	Yes	Yes	Yes	Yes, if correctly applied	Yes	No	Probably not
Reduce public transport costs	No	No	Yes	Yes	Yes	Yes	Yes	Yes	No	Yes	Yes	Yes
Create new industrial jobs	No	Yes	Yes	Yes	Yes	Yes	Yes	Yes	Yes	Yes	No	Yes
Support decentralized economic structures	No	No	No	Yes	Yes	Yes	Yes	Yes	Yes	Yes	No	No
Increase energy independence (industrial countries)	No	Yes	Yes	Yes	Yes	Yes	Yes	Yes	Yes	Yes	Yes	Yes
Increase energy independence (developing countries)	No	No	No	Yes	Yes	Yes	Yes	Yes	Yes	Yes	Yes	Yes

at these cost factors is revealing since it shows the surprising possibility that solar energy actually offers economic advantages!

Economies of Scale do not Apply to Solar Energy Utilization

The law of the economies of scale postulates that by concentrating energy conversion in large facilities, the costs of the various contributing elements (plant infrastructure, management) decline and thus so do overall costs. For example, power generation is cheaper in a few large coal-fired power plants than in many small ones, a phenomenon that produced a steady trend towards large power stations. Centralization does cause additional costs as well – in distribution, for instance – but these costs are overshadowed by the advantages of declining production costs. A precondition is the ability to transport large amounts of primary energy to a central production site – more precisely, those primary energies characterized by high energy density, or fuels with a great energy potential concentrated in a relatively small volume. In contrast, solar energy's widely dispersed, low energy density is regarded as that energy vector's principal economic handicap.

This view ignores the fact that economies of scale apply only partially in the conversion of solar primary energy to a secondary energy form. Economies of scale are valid for large-scale solar energy technologies, but not for the use of these technologies. The reasons have been described by Barry Commoner thus:

> Since the sunlight falls everywhere, an installation is enlarged simply by expanding the area over which the light is received – whether by mirrors for a central power boiler, by collectors for a space heat system, by photovoltaic cells for an electric power system, or by corn plants for producing grain. Each mirror, solar collector, photovoltaic cell, or corn plant is as efficient as the next one. Therefore, a large assemblage of such units is just as efficient as a small one; there is no economy of scale. As a result, a large, centralized solar plant produces energy no more cheaply than a small one (apart from minor savings in maintenance costs). Thus,

with transmission costs eliminated or greatly reduced, the small plant is by far the more efficient way to deliver the energy. This means that the present pattern of building huge, centralized power stations is inherently uneconomical if it is applied to solar energy. No future technical 'breakthrough' can overcome that fact.[146]

There are differences among some solar energy carriers, however, that modify this aspect. Cogeneration plants for solar thermal heat production for 500 single-family homes are more economical than 500 individual units because heat storage for communal plants is simpler and cheaper. For biogas, a single central plant for a village can be more advantageous than 100 single units. But even these larger units should still be regarded as 'decentralized' when compared with large conventional power plants. For solar thermal power plants, spatial concentration of its collectors is more economical, not because of increased collector efficiency but because large turbines and radiation collectors function more efficiently than smaller ones. But turbines over 300 MW offer no more cost-reducing advantages.

But the law of economies of scale simply does not apply to any form of direct conversion of sunlight or wind into electric power or in the conversion of solar power into heat. Large facilities make the use of solar power more expensive than necessary. The logical conclusion is that solar power's economically viable use requires other mechanisms for production and distribution than the power industry. Decentralized planning and construction of energy facilities remains viable without suffering economic losses, and it can avoid the costs of a nationwide, centralized distribution grid. It still remains the task of a genuinely problem- and technology-oriented science of economics to quantify this essentially unassailable basic assumption of solar energy.

Energy Efficiency

Energy efficiency generally refers to the effectiveness of energy conversion in relation to the amount of primary or secondary energy. In such discussions, the 'low' efficiency of the, still young, solar technologies is usually described as a disadvantage. One factor is frequently overlooked, however: solar energy has the unique advantage that it can be tailored to any purpose with the highest efficiency imaginable. The inefficiency of traditional power plants is due mainly to the fact that they require high temperatures for the generation of secondary energy which (again quoting Commoner)

> is produced at an unnecessarily high source temperature and then applied, inefficiently, to processes that require a lower quality of energy. Conventional energy sources usually operate, so to speak, downward in quality and therefore, in thermodynamic terms, are often inefficiently coupled to the tasks to which they are applied.

He adds:

> Solar energy is intrinsically of very high quality and can readily be applied to a task that requires high-quality energy. Solar energy is therefore thermodynamically suited to any energy-requiring task, and can substitute for the present sources of energy in any of their uses.
>
> The reason for this surprising situation is that the thermodynamic quality of radiant energy is determined by the temperature of the source that emits it. In this case, the source is the luminescent surface of the sun which has a temperature of about 10,000 °F. Solar radiation is therefore inherently of very high thermodynamic quality. The low temperature that direct sunlight produces when it is absorbed at the earth's surface (about 100–120 °F) does not mean that the quality of the energy has been degraded en route. Rather, it signifies that the energy has spread out enormously in its long radial journey from the sun ... All that is required to deliver solar energy at any desired temperature, up to the 10,000 °F of the solar source, is to concentrate it from a sufficiently large area.[147]

Solar energy use then does not need to rely on the inefficient production of surplus heat, which can be used only partially as waste heat. With traditional plants, an increase in efficiency means a reduction of energy losses; with solar energy, it means an increase in energy gains. Here we find a much larger potential for an increase in efficiency in economic terms, something that will become especially noticeable with the markedly lower, or, rather, eliminated costs for power plant cooling and with the utilization of solar thermal energy for industrial process heat. This last area has hardly been touched on in the discussions of future solar energy structures.

Operational and Management Costs

Operational and management costs – in business parlance, a part of so-called variable costs – refer to costs that depend on market developments as they affect raw energy (oil, gas, coal, uranium) and processed energy (refined fuels, nuclear fuel rods), on waste removal expenditures (spent nuclear fuel rods, oil sludges, coal slag), as well as transport costs, downtime costs and personnel costs. The established energy business can reduce those costs in the face of competition and new processes, but they cannot be avoided completely. With solar energy, on the other hand, these costs can be avoided completely in some cases or largely reduced. The degree of certainty with which they can be calculated is also markedly higher.

Wind and the sun's rays do not cost anything; costs are always incurred by the facilities involved. Biological waste is also free, or even produces some income for the operator. And, except for some types of biomass utilization (energy plants), the energy does not have to be processed. Waste removal costs do not occur. Downtime costs are insignificant; if a single wind power plant breaks down, others continue to operate. In the case of solar cell and collector assemblies, a few modules may break down, but only seldom does the entire plant shut down – and then only for a short time. Repairs and maintenance work can be completed quickly. Transport costs for primary energy in decentralized biomass plants are low because, typically, only short distances are involved. Apart from imported solar electricity, even electric power distribution costs can be significantly reduced; only hydrogen has transport costs similar to those of natural gas. It is precisely those transport costs that represent a considerable economic burden and, in terms of the national economy, a damaging 'redundancy'. The energy generated in large power plants is transported three times, causing higher energy losses and higher costs to the consumer.[148]

Especially noteworthy is the fact that solar power permits, in part at least, the elimination of some routine personnel costs. If the home owner has a solar installation on the facade or the roof, or if the farmer operates a wind or biogas plant, then their operation is part and parcel of all other activities. Only occasionally will they have to pay for maintenance and repair costs – akin to calling the plumber or other craftsman. No utility has that option, even if it operates facilities of this type. And since the energy efficiency of decentralized facilities is, as has been shown, no smaller than that of large facilities, independently operated generating plants cannot be muscled out by large utilities provided they are allowed to feed their product at fair prices into the grid. They operate more cheaply – a reversal of all traditional experiences.

6

The Nonsense about an Energy Consensus: The Resistance to Solar Energy

A question asked with increasing frequency is: why are all governments and the private sector not seizing these opportunities with both hands, even if they do not yet quite live up to the promise outlined here? It also seems unimaginable that only a tiny number of scientists, including economists, are capable of recognizing these opportunities. Typically the optimistic, yet realistic, arguments on behalf of solar energy are still frequently met with a mixture of incredulous amazement and doubt. There simply has to be a catch with all this, the doubters are saying. The historic frequency with which collective stupidities have been committed by previous generations of political leaders has already been described. History provides ample proof that new theories, superior to past assumptions, have great difficulty in gaining acceptance. Max Planck concluded in his scientific autobiography:

> As a rule, a new scientific truth does not triumph by convincing its opponents and making them see the light, but rather because its opponents eventually die out and a new generation grows up familiar with it.[149]

Evidently the only people able to recognize the profound opportunities offered by solar energy are those capable of dissociating themselves from established ways of thinking. The resistance to new ideas, widely present even in science, has been described by the philosopher of science Thomas Kuhn in his observations on *The Structure of Scientific Revolutions* thus:

> The fundamental rethinking of a conventional paradigm regularly occurs only when it becomes obvious that it is no longer possible to solve a problem with the traditional scientific way and when awareness of that impossibility creates a new paradigm, [an] alternative candidate.

Even then, that process of substitution is anything but rapid and smooth. For the established experts:

> The new theory means a change of the rules that until then controlled normal, scientific practice. Inevitably, this affects large scientific projects that have already been successfully concluded.[150]

In other words, their labours are in danger of becoming waste paper. Not only is their own scientific reputation suddenly at stake, but also the life's work of one or several generations of scientists, and frequently large research resources and funds – money.

This is crucial for understanding why technological and scientific research may be resisting more stubbornly than ever before new priorities outside their own sphere of competence. This is why emancipation from traditional thinking is even more difficult today than when Kuhn described it more than a decade ago. Solar energy will prevail against all opposition because its fundamental advantages cannot be suppressed in the long run, but this optimism is not a good enough reason to moderate our demands and expectations, because at stake is the fateful question of whether this new energy will arrive in time and to the extent necessary to avert the socio-economic and social-environmental dangers looming on the horizon.

Those attitudes, so noticeable in the sciences, are hardly different in politics, business and the media, especially if massive business and national economic interests as well as power politics are at stake – as exemplified by energy supply issues. An increasingly positive response to solar energy is noticeable among some of the media, but it has hardly advanced in political and business coverage. The representatives of establishment energy politics always pretend they are the very incarnation of objectivity when it comes to renewable energy, but in reality they mean their own, one-sided, perspective. If there are conflicts, they call for 'energy consensus' and accuse others of ideological ignorance. They yearn for the past when energy policy and the energy industry were not the dominant themes of domestic political controversy, and the only issues that mattered were guarantees of energy security and low energy prices.

The lack of commitment to solar energy, noticeable at all levels of the establishment, reveals an irrational attitude to existential questions, and mirrors the pervasive helplessness of our present culture in response to the new challenges to our civilization. To overcome these obstacles, we must analyse why and how they retain their hold.

The Conservative Structure of Economic Obstacles

The Solar Conspiracy was the title of a book published in 1975, with the subtitle *The $3,000,000,000,000 game plan of the energy barons' shadow government.*[151] The author, John Keyes, at the time Chairman of the International Solarthermics Corporation, described the futile attempts of small companies offering solar-thermal equipment for home heating to enter and grow in the market. In economic terms, they had all the advantages: plants that would pay for themselves after a reasonable period of time; a price clearly more cost-effective than that of comparable equipment produced by large corporations; and convincing forecasts that there was a large potential market in the offing, larger than the combined sales of all American oil multi-nationals. On the other side, there was a government with an introduction programme that allocated most of its funds to NASA, which in turn was introducing this type of equipment at prices more than ten times higher, and the large companies in the energy business who bought up patents and licences from smaller companies and stashed them away in their corporate safes. The author's conclusion, based on intimate practical experience, was that it was more important to the energy barons to maintain control of the energy system than to earn considerable extra profits from the production and sale of solar plants that would also make their individual owner/operators more independent.

Berman and O'Connor describe, in a book aptly entitled *Who Owns the Sun?*, the decades of struggle in the United States between the super-regional energy monopolies on the one hand and the local city and independent utilities on the other. The efforts of the latter to develop a decentralized and democratically controlled energy supply structure have been repeatedly destroyed by the endless manoeuvring of large-scale energy concerns, more often than not tacitly backed by governments and a mass media that has been bought off.[152] This American paradigm is present everywhere. In Germany it can be seen in the history of the RWE group written by Lutz Mez[153]. There are RWE look-alikes everywhere else too. This is surely one of the key reasons why solar energy has been unable to gain a real foothold in the energy market so far, even when it shows undeniable advantages, as illustrated by these solar thermal domestic facilities. But it is by no means the only motive.

More important is the clear expectation of the consequences of a large-scale shift to solar energy: the most far-reaching structural economic change that has ever occurred! Every change in the structure of an economy produces both winners and losers. At the outset, it is seldom clear who will win or lose. In the case of the large-scale introduction of solar energy, the losers are already defined: they are, above all, the suppliers of the primary energies of oil, coal, gas and nuclear fuels, and, secondly, the traditional manufacturers of large energy-conversion plants and equipment, as well as the owners and operators of these plants. They will:

- lose sales if demand for primary energy declines because of the introduction of solar energy;

- no longer be able to write off fully investments that have already been paid for or, rather, will not be able to continue to use them profitably after they have been fully written off, because the utilization of conventional energy will decrease owing to the expanding supply of solar electricity;

- be less and less in demand in their traditional energy sector because their power plant technology will no longer be needed.

The respective dependences of the various traditional players on the existing energy structures differ widely: some are trapped in a life-or-death relationship, while for others the links are more tenuous. But all would lose a great deal if solar energy were introduced into the market promptly and immediately, and once investments and markets were no longer controlled by the energy industry's old guard.

Beyond this are the numerous interlocking arrangements in the energy business, which means that some companies would be immediately threatened by all of these potential dangers. A significant number of these companies are state owned. These governments thus both have an economic interest in their corporate sales and also depend on high tax revenues from the consumption of these energies. In addition, there is the close interweaving of staff between political institutions and the energy sector, which because of the share of state ownership in energy companies is more prevalent than in other sectors of the economy. The appointment of government or local government officials and civil servants to managing or supervisory boards is a widespread practice, which is always accompanied by a rise in salary. Many are reluctant to spoil their chances of such privileges by stirring up trouble with the energy sector. This is a particular form of legalized corruption along the lines of 'You will get your reward later'. Government representatives and boards of energy companies are often bound together in a 'cosa nostre'.

Oil-producing countries have stakes in vehicle manufacturers; oil multinationals have stakes in the nuclear technology industry, which in turn usually controls the production of nuclear fuels; power-generating utilities own oil corporations and shares of coal mines; and banks own shares of everything. In particular, the big banks, which are natural business partners for the large power companies, have no interest in a change in energy. The amount of lending is too large and the investments are too low-risk, as long as the politico-energy complex remains undisturbed by alternatives. They are the ones pulling the strings and therefore the 'managers of the climatic catastrophe'.[154] These considerations, together with the fact that all these entities are part of an internationally linked, interdependent structure, give an approximate image of the indirect and direct influence and power of this network. Penetration of solar energy into the market is resisted because it would directly and negatively affect the utilization level of production facilities and sales of traditional products. Not only would such a change destroy the accepted projections of an entire, major sector of industry but it would also mean the loss of a large share of capital already invested.

The estimated cumulative world energy investments between 1980 and 2000 were contrasted in a 1980 study that also distinguished between a 'low scenario' and a 'high scenario'.[155] Even with the lower scenario, Table 18 indicates total investments of more than $10 billion: about 60% in the electrical industry alone. This represents an average of about $500 billion annually (at 1980 prices); at current prices, this would be about $900 billion per year. Since these investments would occur, and have occurred, almost exclusively in conventional energy supply structures, these figures give an impression of the magnitude of these investment mountains and, with it, the massive vested interests opposing the introduction of solar energy.

Table 18 *World energy investments 1980–2000*

	Oil	Natural gas	Coal	Electricity	Total
Low scenario (billion dollars)					
Production	1,519	455	353	2,963	5,290
Transport and distribution	585	558	100	3,679	4,920
Total	2,104	1,013	453	6,642	10,212
High scenario (billion dollars)					
Production	1,798	580	708	4,521	7,607
Transport and distribution	612	832	226	5,416	7,086
Total	2,410	1,412	934	9,937	14,639

It is also already possible to name the potential winners from a solar strategy. They are the manufacturers of solar energy conversion technology and individual, local or regional operators. The key difference between potential losers and winners is that the losers are present now, and exert influence. After all, the energy industry in the Western industrial countries is, in terms of sales, the largest and most internationalized industry in the world economy. On the other hand, the potential winners currently are either small, solar-technological enterprises, poorly supported, small R&D departments within big energy corporations, or they exist only on paper.

It is most likely that the main losers will be the producing countries and corporations that market energy raw materials:

commercial primary energy (oil, coal, natural gas, nuclear fuels) will be replaced by solar power, which cannot be commercialized as a primary source of energy. The sun's rays and the wind are free for the operator of appropriate facilities. No company will be able to privatize the sun or to acquire production rights for harvesting it. It is true that the cultivation of energy crops for biomass via the acquisition of appropriate agricultural acreage falls into the area of commercial activities, but those who cultivate them are hardly of the same type as the producers of traditional primary energy.

The suppliers of conventional primary energy have invested large sums for the very long term in the exploration of new regions for production, mining and drilling techniques, in pipelines and bulk carriers for energy transport, and in refineries or processes for upgrading coal. The more and the longer the energy industry has invested, the tougher and the more persistent its defensive attitude is likely to be towards solar energy and even energy-savings strategies. The suppliers in the international energy markets are simply not willing to stand idly by while the revenues to which they have become accustomed, and on which they depend, continue to fall because of declining demand.

Experience suggests that the suppliers will, in keeping with the rules of the market, counter this trend by offering lower prices for their products in an attempt to forestall earnings losses by stimulating demand. In fact, the alternative of price increases would be generally desirable and in the producing countries' own interest because their non-renewable resources would be conserved; but it would have the negative economic consequence of further reducing demand in the short term. In an analysis of the likely reactions of energy suppliers to an energy tax, for example, based on their past behaviour, Mohssen Massarrat, a political scientist, concludes that lowering oil prices could 'destroy with one stroke all ecologically positive effects of a global energy tax, likely to be achieved only with long, difficult negotiations'.[156]

In fact oil-producing countries have always forcefully intervened, through OPEC, against initiatives to launch an energy tax. If they decide to swamp world markets with cheap energy as a reaction to such taxes, a few governments may try to counteract low prices with increased energy import duties or taxes, but most governments will probably swallow the bait of short-term relief for their national economies. In the 1980s, most governments not only succumbed to the lure of lower prices, but also tried to encourage them by attacking the united OPEC front and making agreements with increasingly cash-starved producing countries to increase output and to induce dumping. Apart from Denmark, no other countries tried in the 1980s to compensate for the effect of price reductions by increasing crude oil taxes, which would have been the only sensible reaction. There is little reason to believe that oil-producing countries and oil multinationals would not use this weapon again in the event of a new drop in demand. As long as conventional energy carriers remain the main source of energy, the behaviour of energy consumers can be manipulated by the energy suppliers, who, because of their long-term investments in developing new energy sources, have a never-ending need for revenues. Only solar energy structures would be immune to such counterstrategies, once they were established.

Just as strongly dependent on the existing energy supply structure are the producers of conventional energy equipment. Even if they are not directly involved in the production or marketing of fuels, their product lines are dependent on conventional energy and, correspondingly, their capital is tied to it. In addition, a substantial part of their activities consists of plant repair and maintenance and the sale of replacement and auxiliary equipment. Their know-how and market share are in the conventional energy equipment arena, and they are familiar and comfortable in that territory. Solar energy technology expertise is generally the province of others. The established firms would have to acquire this from scratch, and would have to start again to find new markets for themselves.

Particularly noticeable is the resistance of the nuclear industry. As late as the early 1980s, it was gearing up for what seemed a large, worldwide expansion, which promised to be the deal of the millennium. Around the globe, fewer than ten manufacturers of nuclear power plants had hoped, and still hope, to carve up among themselves a market of hundreds of new plants at prices of several billion dollars each. Nuclear plant builders depend exclusively on a few large customers; in the event of a switch to solar technologies they would find the change to dealing with many small customers especially difficult.

The main reason for the restraint of conventional plant operators has to do with their existing investments. Germany's publicly owned electric utility sector alone made investments of about $50 billion for the five years between 1991 and 1995. Almost all of this was in the area of conventional energies. All new investment in the construction of conventional power plants and new distribution grids represents additional economic obstacles to the introduction of solar energy.

A specific write-off period is estimated for investments in a conventional energy carrier, and the plant's average operating life is calculated. The biggest profit is earned on older plants that are free of debt. Assuming an average of 30–40 years for the operating life of such power plants, operators will not be interested in investing in competing solar energy capacities within that time frame. The more recent such an investment, the smaller will be any such interest, and the larger the burden of old investment standing in the way of new, solar energy investment. This obstacle disappears only when additional new capacity is needed, or when worn-out power plants must be replaced.

The general public should not have to wait until all this investment is written off and profit expectations have been fulfilled. Since each new investment in non-solar power plant facilities is another obstacle for decades to the introduction of solar energy installation, the inescapable conclusion is that any such addition to or replacement of worn-out power plant must be stopped politically. It should also be remembered that individual energy users create obstacles for themselves as well if, for example, they build or renovate houses without sufficient insulation, or if they install new conventional heating or air conditioning equipment. This market is likely to be off limits to solar energy for a long time to come.

These then are the real arguments put forward as to why large-scale solar energy investment allegedly 'does not add up', as its opponents keep emphasizing. This is why it is repeatedly claimed that solar energy represents a viable alternative only 'in the long run'; in the short term, the opposition says, it would disrupt existing investment cycles. Viewed from within their respective business sectors, these economic considerations are undoubtedly logical, and they represent normal entrepreneurial behaviour. However, given the dangers to nature from conventional energy, they are intolerable from the point of view of the common good. The fate of human civilization should not be allowed to depend on the economic efficiency calculations of the energy industry.

Naturally, energy corporations, whether producers, equipment manufacturers or operators, do have the option to diversify their activities: that is, to switch from conventional to solar energy. If, however, they are prepared to switch to a new energy carrier that would collide with their investment plans, they are not disposed to act with sufficient haste.

They are therefore focusing their diversification activities on non-competing investments instead, either within the energy business itself or elsewhere. An additional factor is that one sector of the energy industry's willingness to adapt depends partially on a corresponding willingness to innovate in another sector. A manufacturer of energy facilities will not launch a large production facility for solar plant without clear interest from several operators, and an operator will not entertain ideas of large-scale investment in solar energy while there is no supplier offering volume-produced equipment. It is no accident that the greatest degree of flexibility in entering the solar energy arena has been shown by individual operators who are relatively independent of the system: private individuals, operator cooperatives for wind energy utilization, or independent local utilities in towns such as Saarbrücken, Schwäbisch-Hall or Rottweil in Germany, which purchase some locally generated electricity from renewables or cogeneration plants.

Even if large energy corporations were not subject to these economic constraints, one should not count on any great enthusiasm for solar innovation. These corporations would then have to be willing to turn over a large part of their centralized activities to smaller, decentralized operations, and either break up their own corporation or reorganize into decentralized, independently operating, horizontally structured regional departments. During the transition period they would still have to provide conventionally generated power from central plants to areas without a solar-based power generation system, but this would shrink each year instead of growing. In the end, the centralized services would be limited to operating a grid and handling the import of the fraction of solar energy fed from distant large-scale solar power plants. Any attempt to provide solar-generated energy from central facilities would reduce the economic efficiency, as shown in Chapter 5, and thus slow down its rapid introduction. Large energy corporations therefore have little real incentive to shift rapidly and appropriately to solar energy, but simply stand in each other's way, supported by the credit institutions who want to get their money back.

Since solar energy represents, in the long run, the death sentence for large facilities, it is naive to imagine that major solar-based innovation will get under way as soon as performance and cost analyses have established that solar energy techniques are competitive. The fact that the economics of solar energy are permanently, wilfully and consciously miscalculated, despite all the evidence of its success, is obviously due to the influence of special interests. Once the real economic opportunities and advantages are properly laid out, any publicly supportable arguments in favour of postponing the large-scale launch of solar energy utilization will vanish. In order to nip such shoots in the bud, public enthusiasm is constantly dampened, either with misleading figures or even with absurd statements about solar energy's alleged environmental burdens. With the help of a few small showpiece projects, the opponents of solar technology are feigning effort and interest and constructing a false aura of earnest competence, creating an opportunity to bad-mouth solar energy publicly.

As recently as the summer of 1992, the Bayernwerke utility in Munich began operating a photovoltaic demonstration plant in a village not connected to the grid, for which they officially announced a price of $6.25/kWh, more than five times the cost at other existing German photovoltaic plants. Germany's large electric utilities spend many millions of dollars a year on advertising campaigns against solar energy; this is certainly more than they have invested at their own expense so far in photovoltaics or wind power plants, as they allege they could not justify these costs to their customers. This is in hypocritical contrast to their claim in full-page newspaper advertisements that 'We, Germany's electricity producers, also regard these types of energy positively. And we invest considerable amounts in their development. But let's be realistic: solar power is fair-weather power'.

Meanwhile, renewable energy costs appear to be calculated correctly in one breath and miscalculated in the next. For example, the Union of German Electrical Utilities claimed in connection with the conflict over the German Electricity Feed Law that the equilibrium price of 10 cents/kWh for wind power was unfair and would lead to windfall profits for the wind power companies. However, the member companies of this same organization charge their own electricity customers an inflated 'green tariff' of 18.75 cents/kWh for electricity derived from wind power that they produce themselves. They justify this by claiming that production costs are higher.

A rapid switch to solar energy would demand a degree of entrepreneurial unselfishness by the main actors in the energy industry without precedent in the history of commerce, and an innovative farsightedness that is extremely rare, especially in large corporate bureaucracies. To depend on the energy industry in the switch to solar energy would be to leave mankind to its doom. It would also be highly inappropriate politically to entrust the fate of man to the goodwill and business decisions of a single industry. Were it not for the acute dangers of growing climate anomalies and the critical race against time, a gradual, orderly, well-cushioned transition to solar energy utilization that would take into account the economic problems resulting from the existing burden of investments by the energy industry might be possible. However, the increasing gravity of global environmental danger just does not allow such creeping structural change.

It is irresponsible to let the convenience of the dominant players in the energy industry determine the timing for the general introduction of solar energy technologies, especially as there will be no single best date, given the different energy investment cycles. To seek consensus for structural change in the energy sector simply means increasingly painful delay. 'Energy consensus' is a euphemism for prolonging as much as possible the use of conventional energy and keeping out renewable energy. For solar energy to prevail against the obstinate opposition and disinformation of the ruling energy system calls for an energy battle. Without the long fight by the environmental movement against nuclear power many more nuclear power plants, with all their attendant burdens on the common good, would be operating now.

The array of obstacles available to the vendors of primary energy has been amply demonstrated by their counterstrategies against energy conservation initiatives in some countries, where political efforts to make energy more expensive have been subverted by suppliers lowering energy prices, with both operators and equipment suppliers fearing a loss of turnover. However, their attitudes have begun to change in recent years, in part because of changes in the political constellation, but also partly voluntarily. In particular, some American companies have calculated that the introduction of energy conservation techniques and the sale of energy from performance-optimized plants is better for their corporate balance sheets than investments in new power plants and equipment with long write-off periods. Plant constructors have begun to realize that there is a large and non-controversial demand for components that improve plant efficiency and are not subject to the interruptions caused by civic protests. There is apparently no longer a fundamental conflict, except with energy suppliers, about energy conservation strategies, which is why these are always listed as the main priority in official proposals for energy reform, while solar energy is still relegated to the sidelines.

The opposition to solar energy is much more stubborn: production countries and distributors can influence the pace of conservation measures for conventional energy sources through pricing strategies. Plant and equipment builders continue to work in their accustomed profession and have no fear of new competition. While sales may be affected, the structural influence of the central operating companies is, in principle, not in question. However, as has been pointed out, the manipulation by fuel suppliers no longer works in the face of solar energy carriers: an alternative that cannot be subverted. As solar energy technology involves a much more fundamental conversion of the plant and equipment industry and a fundamental structural change to the operators, it has to be prepared to deal with much more opposition. However, once a real, broad introduction is under way in earnest, it will be unstoppable as it is inherently more resistant to counter-attacks from the established energy industry. Solar energy offers the prospect of getting rid of the influence of these powers.

This economic resistance is based, therefore, on hard-nosed interests and concerns from which the managements of the companies affected can hardly escape. Obviously, were the energy industry more future-oriented, it could have done a great deal more for solar energy than it has so far, but as long as the preponderance of its investments involves conventional energy carriers, it is tied to the traditional structures with lead weights. Although its real room for manoeuvre is considerably larger than it ever admits, it is still far from sufficient, even if fully exploited, when measured against actual needs. None of these companies can afford to discount existing, recent investments and to destroy working capital; none will voluntarily want to be, or can afford to be, the loser in the necessary structural change. Their creditors, if nobody else, would try to stop any such attempt.

To take the easy way out with an 'all winners' approach to avoid hurting anybody would be equivalent to an irresponsibly slow shuffle into a solar energy economy. It is not the allegedly excessive cost of solar energy technology that is the real problem; it is the costs of a prompt and rapid structural change of the energy business that are too high. The energy industry simply represents the interests of a few individual corporations, and not those of the general public. Publicly, understandably from their perspective, they claim to represent disinterested expertise rather than their own interests.

What is not understandable is that politicians, who should be responsible to the general public, and the media generally go along with this charade, which reaches a sort of climax every three years with the conferences of the so-called 'World Energy Council'. This World Energy Council is nothing more than the assembled international energy cartel, to which entire armies of politicians and journalists make their pilgrimages. There, the attempt is regularly made to show where the 'factual' limits of new goals for energy and environmental policies lie. In 1989 in Montreal (at that time the cartel meeting still called itself the 'World Energy Conference') it was determined that solar energy could account for a maximum of only 3% of the world's energy supply by the year 2020. In 1992 in Madrid it was claimed that renewable energy would not play any significant role in the next century (!). Three months after the Rio Conference came a supercilious declaration that annual CO_2 emissions could only be reduced by 10% by the year 2020. What is presented as a forecast is in reality a special-interests decree.

What is needed to achieve the giant steps forward towards solar energy utilization is a consistent political strategy. Rather than constantly trying for a consensus among all interests in the energy industry, which is obviously nonsense for a serious solar strategy, such a strategy must attempt to split up the energy industry! If one looks at the different motives for clinging to established structures, it is evident that plant and equipment manufacturers, regional utilities, new private companies, and other types of business ventures among solar energy producers and users should have the greatest interest in a solar strategy. They are the potential winners whose interests should and must be made to coincide with those of the general public. They need political back-up for an anti-monopolistic, anti-fossil and anti-nuclear stance for as long as they are not strong enough on their own in the battle of economic interests. Without protection, the market power of the energy industry, busy pushing its conventional energy sources, will equate to solar energy's failure in the market place for an irresponsibly long period of time.

Government Failure: The Inability to Influence Structures

The shabby engagement of political institutions on behalf of solar energy is not explained solely by the fact that states and local authorities themselves own large slices of the energy industry and occasionally act more like representatives of corporate interest than of the general public. Nor is it correct to say that all political decision-makers still entertain unequivocal objections to solar energy. Instead, some argue – as did an American administrator at a hearing in the US Senate on solar energy in April 1990 on the occasion of 'Earth Day' – that there is no social consensus for decisive measures in support of solar energy. My response at the hearing was:

> If a government doesn't want to do something, it cites the allegedly non-existing consensus. If it wants to do something, such as armament projects, it pushes through a decision even without such a consensus and calls it 'leadership'.

The failure of political institutions in their solar duty is only one element in the more general failure of governments when it comes to challenges for the future. This is nothing more than the expression of the substantial deficit of today's political systems, which find themselves in a situation of 'a crisis without an alternative', as the German historian Christian Meier described the self-destruction of the Roman Empire:

> Society is enmeshed in structures that advance the crisis process almost out of necessity via the unintended side effects of its actions. On the other hand, it is not yet possible to form new structures that would allow the needy to push forward a new order, for example. There is still general satisfaction with the existing situation, or one is powerless and unable to challenge it, regardless of how little it really works. In this sense there is a discrepancy between that which society wants and that which is really achieved. As long as this situation continues, there is no real force that can confront it, that can translate the crisis into opposing positions and turn it into the subject of political action. In short: so far there is no alternative.[157]

The parallel with the current conditions is obvious.

Government and administrative structures, as they have evolved in recent decades, obstruct each other in their attempts to rise to the challenges of the future. Mirroring the increasingly complex and multilayered structures of the entire society, they have increasingly assumed forms that are dictated by the principles of the division of labour. The cabinets of the 19th century included only the 'classic ministries'; today they have expanded almost everywhere with specialized departments, growing in some countries to as many as twenty or thirty ministries with specialized sections, including recently the environment. Additionally, numerous new government offices and agencies have been created. Parliaments have followed this trend as well. The prototypical modern political engineer is identified by a specialization, so that today there are hardly any examples left of the politician as such. Instead, there are specialist politicians for foreign policy, for finance, economics, transport, education, agriculture, and so on, up to energy politicians and environmental politicians. Behind this is the idea that the architecture of our political systems and their institutions is basically complete. What is left is renovation, modification and addition to parts of the overall edifice. Such models for political action and careers may be suitable for relatively stable periods, but in times of sweeping structural crises they will inevitably fail. Those individuals integrated into the system are capable of only partial change, but are unable to come up with broad new political designs that transcend the limits of individual ministries and departments. Given these limitations, the officially elected, politically responsible representatives become increasingly dependent on the numerically far superior, unelected professional specialists in departmental and ministerial administrations.[158]

The environmental crisis has shoved under our nose the fact that we are part of a total organism and that we therefore have to think and act in terms of the whole. This is diametrically opposed to the modus operandi of present-day political institutions. One department or political section no longer knows much about what others are doing. The actions of one section are frequently negated by the goals of another. Environmental questions cut across nearly all other issues, but in the modern political division of labour, the environment appears merely as one department alongside all the others. This explains why political decisions are made with increasing frequency that seem perfectly rational for their particular sector but which have disastrous consequences in other areas. The division-of-labour methods in modern management, as common in ministerial as in corporate bureaucracies, include mechanisms such as the delegation of tasks that dismember politics further and contribute to an increasingly microscopic way of looking at things. Such delegation is widely regarded as proof of modern political and economic leadership, but the further it spreads, the more political leadership will become dependent on mere piecemeal labour.

The result is that governments are really ruled by bureaucracy, their permanent institutions, and are increasingly interchangeable. The public is waiting in vain for new grand political designs, even though these almost force themselves onto public attention, such as the shift to solar energy. Political leadership and parliaments inevitably depoliticize themselves in such a situation, because they have no creative ideas of their own that they could push through with their formal decision-making powers. This simultaneously encourages society's growing political apathy by permanently frustrating its expectations. Instead of acting on the insights of the German social philosopher Niklas Luhmann, who urges 'reducing complexity', exactly the opposite happens because of the organizational structure of political institutions and their work patterns: laws and regulations are enacted of constantly growing complexity and confusion, increasing the inability for reform, and increasing disgust with politics at a time when sweeping reform and political engagement should be paramount for the survival of mankind. Increasingly dangerous political deficits are being papered over by flights of rhetoric. Problems are treated with words instead of genuine attempts to solve them. This deception of the public works for a while, particularly with the help of the media, but the result is a growing loss of credibility by political actors, resulting in political apathy and, ultimately, rebellion.

This seems to be the only explanation for contradictions, such as the fact that the German government published the positive results of a government-financed test programme for low-energy homes but at the same time finances new public housing programmes that largely ignore these recommendations. Another example is that the Commission of the European Union adopts a suggestion for an energy tax to reduce energy consumption, but at the same time promotes deregulation of European air traffic aimed at lowering air fares, which would generate extra traffic with a consequent increase in the consumption of climate-damaging energy. Similarly, the German research ministry finances climate and solar energy research but at the same time participates in the development of supersonic aircraft whose deployment would inevitably contribute significantly to the destruction of the earth's atmosphere. In addition, climate conventions are agreed, while simultaneously counter-productive political initiatives are decided upon to lower the price of energy. Indeed, governments almost everywhere try to protect the position of the energy business. In so doing, they openly apply economic planning criteria 'to guarantee the security of investments', while they happily abandon the renewable energies (although these alone meet the demands of the future) to the vagaries of the market, in which they have to compete with conventional energy businesses that for the most part still enjoy legally enshrined privileges. All the skewed positions described in Chapter 2 under the heading 'The Failure of Politics to Meet Future Challenges' are part of this category of grotesquely contradictory government action.

It is no wonder then that governments are constantly blocking with one hand what the other hand is trying to set in motion. The social and environmental costs of past and present energy supply are thus disregarded, as are the broad social advantages of solar energy, even for public construction contracts. Cities with millions of inhabitants are choking on their own waste, and disease is spreading, but more thought is given to the construction of a coal-fired or nuclear power plant than to using their organic waste constituents to produce energy to protect the atmosphere, to improve air quality, to cleanse the water, solve the waste problem and support public hygiene. Just as the energy industry is sticking to its traditional structures and its current investments (for which there exists at least a certain, if limited, economic logic), the system of political institutions and the processes they control cling to deeply rooted modes of operation. They are incapable of taking steps towards and making plans for the future in a way appropriate to the magnitude of the problem while they adhere to the well-trodden paths of political business. It is a truism that, to move to the path towards a global solar energy economy with the necessary rigour, the economic obstacles must be overcome with strategic political decisions. To do this, it is necessary to encourage political institutions through democratic public pressure to move towards a solar strategy. For that to happen, we have to make clear to the general public that this alternative is truly possible, so that increasingly pressing questions are asked about why this opportunity is not being grasped, and so that the demand for a practicable solar strategy turns into an ultimatum.

With the growing disenchantment with politics, it has become fashionable to accuse the current generation of politicians of lacking ability, while pointing to the towering personalities of the past. But this comparison merely veils the real problems and is, despite the current shortcomings of governments, extremely unfair. The present political figures are faced with far more problems to solve than any preceding generation. Environmental problems, excessive armaments, Janus-headed technology projects, orgies of wasted energy, excessive bureaucratization and, last but not least, the entrenched structures of control and decision-making are all problems with their roots in the decisions of the much-praised predecessors, the overwhelming consequences of which can no longer be ignored. Many past political decisions were made blind to their future consequences, which partially explains the current state of political helplessness. Today's generation has to pay for the mistakes of previous generations. It is very difficult to see how these consequences can be corrected in the short term.

The real weakness of most of the current politicians has nothing to do with the charge that they are not as good as their predecessors. It is something different: they cannot formulate a real alternative as long as they are tied by the established patterns of action. Originals are always more impressive than recent pale imitations who may still be suitable for smooth, successful individual leadership careers. The requirements and capabilities currently required to gain political leadership are increasingly in stark contrast to the actual ability needed to handle these responsibilities today. Conventional patterns of behaviour are at best sufficient to wield 'power within existing conditions', but not any more to exercise creative 'power over conditions'. Thus, the 'power to act and impotence to change' exist side by side, forming a deadly mixture because the ideas for a more appropriate contemporary political concept have not yet prevailed and there are simply not enough active reformist participants. The political neglect affecting a central area of future concern, such as solar energy, has a deeper cause; that these omissions continue is also explained by the observation that politics does not occur in a vacuum but is significantly influenced by ideological currents.

What has been described so far in this regard as 'Failure of the State'[159] has structural causes, which do not apply solely to the question of renewable energies. In energy policy in particular there are also some very profane reasons for failure, which we encounter everywhere: the all too intimate relationship of staff in political institutions with those in the energy business – a relationship that runs deeper than that in other sectors. The origins of this lie not only in the fact that large parts of the electricity business began life as public sector enterprises, or are still in the public sector and feel as though they are part of the system of public institutions, with all the concomitant interchange of personnel on management and supervisory boards. Another cause is the – correct – recognition by politicians that the energy supply is the basis for existence of the overall economy. This endows the energy sector, in addition to its actual economic position of power, with a psychological advantage. Because society is dependent on constant availability of energy, political institutions feel dependent on the energy sector. Whenever the latter threatens economic disruption, governments, parliaments and other economic players alike tremble at the knee. The energy business represents in this way a projection of power, which enables it to feel and behave like a state within a state. It is conventional political custom not to interfere with the energy sector.

This is how the pattern of behaviour of political institutions arose, giving the energy business considerably more direct political influence than other economic sectors – comparable only with the direct influence of the arms industry. One can readily observe that the ministries responsible for energy behave like agencies for the sector. Legislative proposals on energy that governments put before parliaments are frequently composed entirely by the energy sector itself. I explained this years ago in Vienna in a public debate with the former Chief Executive of Verbund, the Austrian oil company. He emphatically denied such a description of the government as a puppet of the energy business. However, not half an hour later, his personal vanity betrayed him when he proudly told of the incident when Austrian Chancellor Kreisky once asked him in his role as head of Verbund to write a draft energy supply law. An obvious intimate relationship between politics and the energy sector is shown in the following example: in 1994, under the aegis of the Bavarian Finance Minister, Mr von Waldenfels, the state-owned hydro power stations on the Rhine/Main/Danube canal, which had only been brought into operation in the 1980s and had cost the taxpayer $2.5 billion, were sold at a knock-down price of $438 million to the private company Viag. Only a few months later the Minister appeared on the board of this obviously favoured company. In 1994, the German electricity utilities, headed up by Preussen Elektra, based in Hanover, started the campaign against the Electricity Feed Law for renewable energies and demanded that the supplemental costs be met in future not from the full coffers of the electricity companies, but from the empty coffers of the state; this would surely kill off the budding wind power industry. Precisely the same demand appeared in a draft law put before the Upper House of the German Parliament by the Hanover-based state government of Lower Saxony. Then

Preussen Elektra complained to the EU Competition Commissioner about the Electricity Feed Law using completely inflated claims of supplemental costs. Although the precisely correct calculations lay before the Competition Commissioner, he adopted seamlessly the version given by the electricity concern, supported by the German Economics Ministry. Furthermore, in 1997, when there was a revision to the German Energy Sector Law, the handwriting of the electricity industry could be read between the lines of nearly all the draft proposals emanating from the Economics Ministry. All these examples – and there are comparable ones in every country – underline the need for greater transparency in these proceedings so that the public opposes them and no longer puts up with the excuses given for why there are no alternatives to continuing with activity that results in destruction of the environment.

Soft Errors

It would be wrong to overlook the fact that another kind of objection is raised against solar energy as soon as it ceases to be a marginal source of energy. These objections are based on structural arguments that have a fundamental problem with large-scale alternatives. They come from the conceptual world of the early environmental movement, which pits 'hard large-scale technologies' against 'soft' and 'small-scale' alternatives. They emanate from the tradition of classical nature and landscape preservation, which resists any interference with natural landscapes that have remained hitherto, either actually or apparently, undisturbed by the activities of civilization, even if it is the activity of renewable energies. They represent contradictions that until now were often swept under the carpet in the environmental movement, but which have come to light with the introduction of renewable energies. They must be overcome if the solar energy alternative is to develop fully, even if it leads to a schism in what we think of as the environmental movement.

> The movement first arose from the attempt to prevent widespread environmental destruction wherever it cropped up. However, prevention is no real alternative. If no alternative is offered to follow

plain prevention, the environmental movement will run itself into the ground fighting nuclear and coal-fired power stations. People rally more readily to the cause of 'no' to certain projects than under the banner of 'yes' to alternative projects. However, above all it is not necessarily always the same people who get together in a citizens' initiative to protest against projects or to demonstrate in favour of an alternative. Blatant selfish local motives often lie behind the 'no' protest, with which many people who think in a more cogent fashion cannot identify themselves. A 'yes' initiative is often based on values in which new opportunities for society can be seen, or on self-interest for those who see an economic opportunity for themselves. So far, it is negative thinking that has fashioned environmental and nature protection legislation and with it the behaviour of environmental and nature protection agencies.

When the environmental movement took action against nuclear power or against new fossil energy development sites, the possibility of a comprehensive alternative through renewable energies had not yet been developed. Even today many who think and act in a thoroughly ecologically responsible way in favour of alternatives are not aware of this possibility. While they advocate renewable energies, they question and doubt whether mankind's energy needs can be met in full by them. That is why they adhere to the idea that it is necessary to retain the option of further nuclear and fossil energy use in perpetuity, as Klaus Heinloth does in his book *The Energy Question*.[160]

Other schools of thought concentrate on the option of energy conservation and efficiency. They usually underestimate the limits of this approach outlined in the second chapter. However, unless they are advocates of options such as fast breeder reactors or nuclear fusion, since they cannot see beyond the fact that the supply of fossil energies is exhaustible, they formulate an ethic of abstinence, a sort of energy austerity policy, which must potentially be adopted by the entire world population. In my view this austerity approach has no chance of being adopted on a voluntary basis by world opinion. Lined up against the mass adoption of austerity we have the daily drizzle of product advertising in every nook and cranny, which urges us to consume and thereby to use energy. Anyone can imagine which approach has the greater power to influence the mind. There is not much chance of the austerity approach influencing mass behaviour on a socio-psychological basis. However, with renewable energies we need not exercise abstinence and can still preserve the environment. It comes down to a struggle between those who adhere to the idea of abstinence while remaining ignorant of the possibility of a solar alternative, and those who want to introduce renewable energies on a broad scale. A self-limiting pattern of behaviour collides with a dynamic one even if both are thinking along the lines of responsibility for nature conservation.

Anyone who perceives environmental policy primarily as protection against the industrial and consumer society with the help of targeted restrictions must necessarily come to an ever-increasing tangle of environmental controls that spawn a new bureaucracy and potentially have a crippling effect. It occasionally leads to strange answers: when I suggested the initiative of a global afforestation with regionally typical mixed forest for absorption of CO_2 to members of the German Parliament's Commission of Enquiry for Protection of the Earth and its Atmosphere, one member replied in all seriousness that an environmental tolerance test would have to be conducted first. The imposition of restrictions to protect the environment is in some ways burdensome to the representatives of industries that destroy nature when it affects their projects. In other ways it is highly welcome when it obstructs alternatives. For example, a representative of General Motors seriously objected in public against afforestation proposals on the grounds that we should be careful as we know too little about the emissions from trees. Electrical companies point with glee to protests against wind farms that claim that they would destroy the landscape in order to adhere with all the more conviction to the necessity for further nuclear or fossil energy use. The German Federal Environmental Office published an expert opinion years ago that condemned the use of plant oil in engines because the positive energy contribution of rapeseed oil is only one third, owing to the energy required for cultivating rapeseed. Instead it favoured the use of energy-saving engines, for which small advances of less than one third are welcome. It never occurred to the authors that plant oil motors enable more efficient energy use. The same Federal Office contested in a further expert opinion a connection between the use of small running-water hydro plants in rivers and the target of reducing CO_2 emissions, although such hydro plants undoubtedly produce completely emission-free electricity. Once again they favoured, instead of this, an increase in energy efficiency of traditional power plants, although these never can reach completely emission-free operation because there is no perpetual motion. Nature protection agencies vent their feelings – remember Peter Kafka's remarks about the law of entropy – while offering the operators of renewable energy plants negligible support.

1. The objection that one large-scale technology must not merely be exchanged for another confuses the wrapping with the content. It takes its cue from experience that is simply not applicable to solar energy technology if we examine its specific economics in relation to nature's energy supply. Even if it involves simply an exchange of technologies, the argument is without merit because this exchange means nothing less than the replacement of environmentally destructive with environmentally benign energy sources. The source of energy actually used and the conversion processes are, and will remain, the core problem; everything else is secondary. The size of energy facilities in and of themselves does not represent an existential danger for humanity, and would not have triggered a worldwide protest movement. In a solar power plant, the maximum credible accident would be the failure of a few modules requiring their replacement, a procedure that can be performed immediately, without danger and without having to shut down the entire power plant. The pair of opposites 'soft' and 'hard', or 'large' and 'small', cannot be simply transferred to solar energy. What does 'hard' and 'large' mean in terms of solar energy?

   There is a single solar energy technology, though, where the analogy with large nuclear technology is perhaps appropriate. This technology exists only on paper, and will have to be kept there. This involves solar power plants in space. As early as the 1970s, the American physicist Glaser developed the idea of stationing a solar cell platform measuring 11.7 by 4.3 km

(7.3 by 2.7 miles) in space. The station could continuously convert solar energy into electricity because, except for a few hours every year, it would never be in the earth's shadow. A 10 GW energy stream, equivalent to the output of ten nuclear power plants, would be transmitted with a 1 km (0.6 mile) diameter microwave beam to a 50 km² (19.3 square mile) reception antenna. Since massively bunched microwaves of this type are extremely harmful, and since this type of solar energy, in contrast to insolation naturally occurring on earth, would funnel large additional energy volumes into the earth's atmosphere, such a satellite power plant cannot claim the positive features normally associated with solar energy.[161] Even today, work is continuing on these concepts; as recently as the late 1980s, the German company MBB had plans for a development of this type, but never put this idea into practice.

Thus, let us concentrate simply on solar energy utilization here on earth. In this context, large power plants can only mean large tidal power plants or hydro power plants with huge reservoirs and high dams. There are good reasons for rejecting such facilities as well, just as the cultivation of giant fields of monoculture energy crops with excessive use of fertilizer for furious plant growth and increases in yield should be rejected.

But if these mistakes can be avoided – and, as shown in earlier chapters, they can – then large acreages for harvesting biomass, or large solar collector and solar cell fields, are unproblematic. These facilities are not large-scale technology but large-area technology. Deploying large-area solar power plants would be economically necessary only if the existing total stock of buildings and their surfaces for utilization were insufficient for a solar energy supply system. Realistically, this is not likely to be the case in densely populated countries. Construction of large-area solar power plants would be restricted to unpopulated regions such as deserts and other marginal land, and might lead to their cultivation with the help of irrigation.

Obvious practical and economic reasons will generally militate against locating the entire production of solar energy in distant desert areas and covering tens of thousands of square kilometres, an image that can cause hostile reactions. However, if there is no other option, then one would have to accept this type of monoculture as well. While this would present some problems when contrasted with the supply of energy from fossil or nuclear power plants, the politically relevant comparison, it would be seen as furthering the grand scheme of rescuing the earth from climate and nuclear dangers.

However, the practical priorities of a future solar energy supply rest in its near-term potential, not in the distant future: in the immediate utilization of the domestic energy potential available in every country, and in the wide dispersal of solar cells, solar collectors, wind power, biomass utilization and small-scale hydro power. Distant energy facilities would be needed only to make up for any possible energy shortfall through solar energy imports. These imports would clearly be, in any event, smaller than the current energy purchases of energy-importing countries, especially once the energy efficiency revolution had taken root. The only aspect likely to be large-scale will be the manufacture of these facilities, where the law of economies of scale will apply. This type of large technology will have to be accepted, just as a tenfold scaling-up of bicycle-manufacturing capacities will have to be accepted to handle emerging new patterns of transportation.

What Ernst-Ulrich von Weizsäcker wrote in his book *Earth Politics* in this regard can therefore be rejected:

> With solar, hydro, wind, biomass, and geothermal energy, scarcely any thought has been given to the environmental impact since their contribution is minimal (except for hydro power). One must, however, agree with the IIASA that once these energy sources produce several gigawatt or terawatt they will bring about major environmental problems worldwide. One need only think of the Brazilian dam projects or the massive cultivation of energy crops.[162]

'Major environmental problems' sounds comparable with the damage wreaked by the use of nuclear or fossil energy. This statement cannot be justified. The largest dam in the world, which is currently being built in China, with a planned capacity of 20,000 MW, will lead to the resettlement of more than 1 million people, and huge areas will be flooded.

However problematic this may be, it is still a better alternative than 20,000 MW of capacity of nuclear or coal-fired plants. The environmental impact of this reservoir and dam is regional. It has no global consequences. The same applies to large-scale plantations. I raise this example although I am no fan of such massive projects, because there are as a rule better alternatives within the field of renewable energy which should be preferred. That is why we sometimes refer to new renewable energies, in order to emphasize their difference from renewable energies derived from large reservoirs or the diversion of entire rivers, which is practised in many places. Reservoirs and dams that are already in place should on the whole be left alone, instead of being replaced with other renewable energies, at least for the time being, since the priority must be to substitute for nuclear and fossil energy. The future of further hydro utilization lies in running-water hydro with negligible local impact on the landscape. As far as energy crops go, we must take care in each case that – be they food, energy or raw material plants – they are farmed on an extensive and multi-cultivation basis. When Ernst-Ulrich von Weizsäcker refers to 'all' renewable energy sources leading to major dangers if they were introduced on a very broad scale, we must contradict him. With wind power, photovoltaics, solar collectors and other technologies, notable environmental problems can only occur in their production: no more or less so than with the production of tractors, turbines, cars, glass or silicon chips. However, it would not be necessary to alarm world opinion about the production of turbines or motors. It should be alarmed, however, about the massive application of nuclear or fossil energy used to power such industry. The laws of thermodynamics tell us that the all-important question is the energy source, not its conversion technology. Over and above this we do not dispute that the production of the conversion technology should be achieved as far as possible without harmful side-effects.

2. The argument that is frequently heard, that energy consumption habits and lifestyle must change before the massive use of solar energy, overlooks the relationship between technology utilization and structural development.

The defensive posture towards solar energy among the supporters of conventional energy is, after all, based on the knowledge of exactly the extent to which the deployment of solar energy technologies would alter existing organizational structures towards more decentralization and independence. It would be nonsensical and impossible to seek to change structures for a technology that has not even been introduced yet.

Techniques leave their mark on political and social patterns and, once created, these structures demand their appropriate technologies. Structures created by this interplay will automatically be modified by solar energy once the techniques required have been efficiently deployed. The same is true when it comes to the demand to modify lifestyles. Most people's consciousness is not changed primarily because of their understanding of necessities but, recalling a basic sociological insight by Karl Marx, by changes in their economic circumstances. In other words, it changes with the practical reconfiguration of the energy system, but not before such a change.

3. It is true that even solar energy technologies create environmental burdens. Naturally their installation must proceed with careful regard for the land. Nevertheless, these objections, presented occasionally by some conservationists, pale in comparison with their alternatives. They indicate that the hierarchy of danger is frequently not taken into account in assessing environmental burdens. It is environmentally disproportionate to equate the irreversible dangers for man from traditional energy sources with reversible regional intervention in nature via solar energy utilization. Naturally, a virgin section of coastline is more beautiful, but the choice is not an unspoiled coast versus wind plants, but continued oil spills into the sea, CO_2 emissions, and electricity from nuclear power versus wind power plant. The real equation is that for every 500 kW capacity wind power plant not constructed, 1,000 tons of CO_2 are emitted annually by coal-fired power plants. It is environmentally irresponsible to oppose the construction of a wind power plant because of concern about nature conservation and thus implicitly accept that the destruction

of nature by nuclear or fossil energy cannot be reduced by a corresponding amount. To put it bluntly, there is a tendency to tie the introduction of solar energy to the precondition of absolute nature conservation where not even the trace of a problem remains. But when such projects fail, the process of total destruction continues. If such absolute standards were applied to energy conservation, the imbalance would be much greater since cutting current energy consumption in half would still mean that the other half would continue to be used, with corresponding consequences. This sort of objection can go as far as a general rejection of wind power. For example, the environmental economist Hans-Christoph Binswanger does this, with the consequence that the dangers of conventional energy sources and conventional damage to the land are minimized.[163] Such objections can therefore lead to resistance reminiscent of the protests against nuclear power stations. It is all just so much hot air, for which there is an excellent description in England: BANANAs (build absolutely nothing anywhere never again)

4. Some fear a massive move to solar energy because unexpected, serious consequences may occur in the future. The experience with nuclear power is still vivid in the minds of many: when an entire generation bet in good conscience on this form of energy and only discovered the true consequences later. It is a myth, however, to suppose that the dangers of the civilian use of nuclear energy were not known in the 1950s. There were many discussions about nuclear power in the 1950s and 1960s when the impending dangers were pointed out forcefully and, a few decades later, these predictions turned out to be correct.[164] With solar energy, on the other hand, it is impossible to discern even a hint of any comparable problem. Where could the problem arise, it may well be asked, when the main issue is merely to harvest solar energy, which arrives on earth anyway, by technological means? If there were such a problem, the multitude of special interests opposing solar power, who can afford to buy entire armies of consultants, would have moved in force long ago. Instead, they have to rely on ridiculous, phoney arguments.

Perhaps fear of making a serious new mistake and defensive environmental concepts explain why so many people limit themselves to making minimum demands for energy conservation because they feel this can't go wrong. But this increases the danger of moving in the wrong direction by erring on the side of caution, a stance not very different from the hesitancy of the energy industry, which practises a strategy of procrastination with a few projects here and there instead of promoting a genuine introduction strategy. If only small demonstration projects are advanced, funded by peanuts, instead of a really comprehensive general strategy, no wonder hardly anybody believes solar energy could ever be the mainstay of our energy supply.

The reservations of part of the environmental movement against a broad-based solar energy offensive turn out to be 'soft errors'. Unintentionally they block up the very core of their concerns.

Talk Globally, Postpone Nationally: The Abuse of Internationalism

'Think globally, act locally' – this catchy slogan of the environmental movement urges everyone to play their part in society, acting on his or her sense of responsibility. Internationally, reality works according to another, unspoken formula perhaps best described as 'Talk globally, postpone nationally'. The global dimension of environmental dangers is continuously abused by national governments as an excuse either to water down or to put off completely long-overdue initiatives as long as nobody else acts on them. It has become the favourite game of governments to redirect expectations of them to some other international arena such as a conference or some organization. It is always easy to come up with reasonable-sounding explanations: it won't accomplish much if only one country takes the initiative; since we all compete internationally, all the other countries would have to go along to avoid incurring economic disadvantages by imposing higher energy taxes in one country only.

While the need for global changes is essentially correct, it is very dangerous for the common good to make essential initiatives wholly dependent on such changes. It means burying perspectives for the future, because it is almost impossible to expect all governments to display the same political will and interest at the same time, and to have the same political and economic room to manoeuvre for new strategies.

There is a basic fallacy underlying the countless environmental conferences under the aegis of the UN: that before anything can happen, it is necessary to achieve a general consensus of all governments that reflects all existing power, interest and development differences for such a wide and complex range of problems. Of necessity, United Nations government conferences must take their cue from the consensus principle.

It is no wonder that individual initiatives are destined to be postponed and sidelined without compunction. More astonishing is the degree of political naivety with which governments, environmental groups, international organizations and the media keep projecting great expectations and hopes on such conferences that are impossible to fulfil. For years before the actual event, international organizations and individuals active in environmental issues were fixated on the impending conventions for the protection of the environment announced for Rio de Janeiro – as if there had been no experience with conventions of this kind. At the end of the 1960s, for example, a convention was adopted that specified that every Western industrial country would contribute 0.7% of its gross national product to development aid. In the thirty years that followed, only four countries lived up to that commitment. There are few reasons to believe that anybody will live up to the obligation of the Kyoto protocol from 1997 for saving the climate any better this time. International conferences are used to articulate goodwill as a substitute for failure to do anything, and in an attempt to mollify the public. If they do not produce any real action, it will be blamed on the complexities of international understanding. At least, it is usually emphasized, such conferences serve to formulate new programmatic levels of understanding that perhaps can't be ignored as easily as before. But these short-term hopes that cannot be met by the parties responsible lead to dangerous disappointments. Left behind is an increasingly discouraged, ever more fatalistic public, more likely to switch off than to become involved.

The international treatment of an energy tax is another case study showing how responsibility is pushed aside. As soon as this demand surfaced in Germany, its basic correctness was endorsed in principle, but at the same time the need for a common international framework was pointed out – at least within the European Union. After the Commission of the European Union came up with a joint position after long debate just before the Rio Conference, the plan was dropped entirely under pressure from some member governments, who pointed to the Rio meeting – once it had become clear that no such resolution would be passed there. The latest excuse is that as a result of increased international competition due to economic globalization, this is no time to be imposing higher energy taxes.

The principle that a basic shift to environmentally safe energy must be agreed on internationally is an excellent vehicle for hiding from a critical public one's own interest in perpetuating the wrong choices. Since not only one government but the entire community of nations finds itself on the same dead end street, and since everybody is using the same smoke-and-mirrors tricks, this maxim is proving to be an extremely cumbersome obstacle in the way of innovations. This is true especially for the global energy system, which is largely international in its make-up and is characterized by the fact that almost everybody concerned has made the same mistakes. It has become hackneyed and a stereotype to claim that any meaningful steps of one's own towards better care for the environment will lead to economic disadvantage if the international community, or economically or politically linked groups of states, do not act in concert.

It has been shown that the economic advantages of solar energy far outweigh the initial difficulties, making it irresponsible to wait until everybody in the international community jumps on the bandwagon. If we look at the difference between centralized and federal – that is, more decentralized – political systems, the latter usually have a few significant advantages. In centralized systems, basic errors percolate through the entire system; in federal structures, parts of the political system can look for new answers that, if successful, can serve as a constructive example for others. Unfortunately, it is exactly in the critical area of how to shape the future where nobody practises anything such as a global federalism – only a stubborn international conformity, which has become even worse since the demise of the Eastern

Bloc. Particularly on questions of environment and energy, it is imperative to find new answers that do not depend on any sort of prior consensus, and to explore one's own geographic options for the exploitation of renewable solar energies at one's own discretion. It is time to replace the environmentally destructive competition among countries with environmentally protective competition among the world's nations.

Another way of sustaining this conformity is the multinational projects that exist in many forms: in space research via the European Space Agency; in the joint development of a fusion reactor under the guidance of the European Union; in joint defence projects such as the Tornado or the Euro-fighter; or in the more recent American-Russian space-based missile defence technology agreements. The official reasons advanced for them include lower cost expectations and the promotion of international cooperation. Whether these cost savings materialize is highly questionable, but another, much desired, effect becomes evident: projects of this type are removed partially or entirely from the scrutiny of democratic decision-making. Anything labelled 'international' must not be questioned domestically because it might affront partner countries. This is an excellent way of removing such projects from criticism and questions, and it is evident that projects are managed this way in areas where the participants wish to avoid critical public discussion.

The wretched process of destroying nature can no longer be halted by creeping long-term strategies, but only by quick, widely effective, here-and-now strategies. The international level of action is relatively poorly suited to the purpose. Most useful, as they have always been, are the institutions of national governments and perhaps soon the joint European institutions within the European Union once they acquire democratic decision-making processes. The revolutionary redirection of the energy supply towards solar energy must include as a prime target the appropriate reshaping of these political institutions. This has nothing to do with nationalism, with any egotistical behaviour by one state at the expense of another. Because of the global nature of environmental destruction via conventional energy consumption, national solar energy programmes would benefit everybody and hurt nobody – except those who trade in the destruction of nature at the expense of mankind. There is not a single reason to be considerate to them any longer.

Nevertheless, there is enough work to do at the international level: not so much the laborious negotiating of non-binding international commitments as the strengthening of internationally effective political institutions that must be able and authorized to act independently, and which must be given the necessary funding. This is needed to enforce international law, and to support those parts of the world that as yet do not have sufficient administrative, scientific-technical, industrial and financial means of their own to do their share for a new politics of the global environment. It is no accident that questions of new institutional political opportunities were treated only in the margins of the Rio Conference.

The Appeal of Traditional Thinking

Those who demand anything new are faced in principle with the need to provide proof: to convince others that the new order will be better than the status quo. This explains why conservatives usually have an easier time pleading their case than innovators. They find support from firmly imprinted patterns of thought and established habits, denounce new approaches with unsupported claims, and create fear and uncertainty about the untried novelty. The more thorough and comprehensive such innovations are, the more there are motives and opportunities for this type of psychological warfare. Solar energy questions so many things that the psychologically understandable attraction of thinking in old patterns becomes – consciously or unconsciously – a real, major obstacle.

In the scientific community, an overwhelming number of nuclear physicists and energy and economic experts confront a much smaller number of solar scientists and environmentally oriented economists. It is no wonder that the former deploy the full weight of their scientific and public stature, earned in decades of scientific work, by looking for cogent arguments in the battle against a new approach they have neglected. Even the great authorities stoop to this level.

A remarkable example is Manfred Eigen, Nobel Prize winner and director of Germany's Max Planck Institute for Biophysical Chemistry. What one can read in his essay, 'Beyond Ideologies and Wishful Thinking', about solar energy is, unfortunately, pure nonsense.[165] He sees, for example, only two means of using solar energy: domestic heat generation and photovoltaics. Many widely recognized specialists of all colours – scientists, architects, department heads in ministries and large corporations, specialist journalists and politicians – would regard their personal and professional reputation as endangered if they had to admit that their special expertise had been off-kilter about a central issue. For this reason they are intent, above all, on proving why certain new ideas just cannot work. On the other side of the coin, it is no accident that most significant technological breakthroughs were not the products of large corporations, and that most political innovations were not the brainchildren of large administrative structures, but came from the outside.

One variant of this problem is giving too much weight to outmoded thinking. In this vein, the renowned German Physics Society published a memorandum in 1995, intending to take a major step to advance the cause of renewable energies. It admitted – and this was (surprisingly) new for many physicists – that renewables could contribute considerably more to energy provision (30%) than had hitherto been widely assumed. I spoke before the society in their headquarters in Berlin, and asked 'why not 100%?' In my speech I indicated the concrete proof of such a possibility without anyone articulating any basic scientific counter-arguments, although I invited them to do so. Apparently, the quantification of the contribution of renewables was a compromise, having taken into account much lower figures from certain members. Even many solar scientists are reluctant to push the theory aggressively and to assert that 100% renewable energy is possible, even though the timeframe for implementation is still debatable. Apparently solar scientists are afraid of being branded as 'lightweight' by those who still think quite differently from them.

Even if it were recognized that a solar strategy is technologically possible, there is a widespread belief that it cannot be achieved organizationally and, with that, sociologically. The result is a sense of despondency, in part self-imposed, in part organized. The fear of new designs, even of those that offer the hope of salvation, engenders the risk of creeping suicide. This is a classic symptom of the decay of a civilization that can probably only be overcome by a generation that has not yet been psychologically exhausted.

7

Return to the Sun:
The Humanization
of the Industrial Revolution

Reason demands that the opportunities indicated here for humanizing the Industrial Revolution by way of solar energy should be seized immediately. The ecological, social and economic state of the world does not permit any further delay to a solar strategy. The goal of building an energy supply system based on solar energy must be the single highest priority in order to achieve the kind of politics and policy that genuinely correspond to the core problems – not only in the individual fields of energy and environmental policy, but in all aspects of economic activity.

To break the ice, there must be crash programmes, and new political and economic players. A solar strategy must be designed for the long term, but that does not mean inactivity should be tolerated in the short term. But above all, the priority of a solar strategy demands that old priorities be dropped. Only then will there be the prerequisite flexibility and freedom of choice.

Time and again there have been examples of political attempts, even in the recent past, to define a new central task and to turn it into reality in a time period that would previously have been regarded as unrealistic. The Manhattan Project for the rapid development of the American nuclear bomb during World War II was one such instance, as was the Apollo Program for an American landing on the moon, the Strategic Defense Initiative (SDI), and France's massive civil and military nuclear programme. All these technical programmes have left lasting marks on the political and economic landscape, and have deeply influenced the value systems of the societies that supported them. It is no accident that all of them were military programmes or, rather, programmes involving the struggle for global power. They do not serve too well as models for a solar strategy, but they nevertheless demonstrate the possibility of organizing political breakthroughs. A better model perhaps is the massive effort during the second half of the 19th century for the construction and operation of the railways, to which Germany allocated as much as 10% of its national product in some years.

It is the peculiar difficulty of a solar strategy that, in contrast to the projects just cited, it cannot rely on a consensus of interests among society's power elites, but will have to prevail against them. Also, it will involve not simply a single large industrial project, but many such new projects in parallel within a common frame of reference, as well as bringing political institutions into line with this task. These institutions possess the formal authority to break down the resistance of interests that are united in damaging the common good. It will be necessary to identify those willing to champion a solar strategy who are capable of achieving a basic behavioural change on part of the political institutions.

What in these structures can overcome resistance against impediments to solar energy; through whom, and how this can be achieved will be described in the following catalogue of action for a solar strategy at national level, while similar measures for a global strategy will be discussed in Chapter 9. The national programme has been so formulated as to make it applicable to any industrial country. The implementation of even a few of the measures described here would lead to breakthroughs in the most diverse areas, all of which taken together would produce a blueprint for saving mankind that could be put into action within a few decades.

The New Maxim: Conventional Energies as Supplementary Energies

To the extent that their existence is acknowledged at all by the energy policy and business establishments, solar energies are regarded as mere 'supplementary' energies. The term implies that solar energy is not a viable replacement for conventional energy carriers, but merely one means of supplying the additional amounts of energy that conventional energy cannot supply. This linguistic usage relegates solar energy to the level of some kind of accessory to the main energy supply, suitable for luxury and secondary demand, but not for 'real' power plants and 'real' energy equipment. If this limited interpretation remains in vogue, the current generation may as well order the coffins for its descendants now.

The energy imperative provides the basis for solar energy's organizing precepts, as formulated by Irm Pontenagel, managing director of EUROSOLAR:

Beginning right now, and not at some fictional future point in time, solar energy must be regarded as the base for energy production. In contrast, the conventional forms of energy must be regarded as additives that may be used only during the time period needed for their total replacement.[166]

This is more than verbal repartee: it is the reversal of the current energy supply's value system, and it must have an action-inducing effect. At first glance it may seem grotesque to regard the energy resources that today represent the essential basis of man's energy supply as mere supplements. But with the basic frame of reference, the earth's natural energy system, classing fossil and nuclear energies as additive does correspond to the true relationships. Accordingly, there must be a rapid reduction in mankind's systemically alienated energy conversion activities in the areas of fossil and nuclear energies (which, together with the forest and vegetation destruction, account for more than 80% of the energy supply) to 70%, 60%, 50%, 40%, 30%, 20%, 10% and finally 0%.

It follows, that, in principle, any use of nuclear or fossil energy should be avoided where, under reasonable conditions, it is already possible to use solar energy. It is no longer acceptable that destructive energy is used because of egotistical interests, thoughtlessness or laziness, if alternatives are available. At the very least, the further use of current energy forms must no longer be permitted as soon as solar energy no longer presents any unmanageable economic and functional disadvantages. This maxim offers a number of starting points for the practical launch of a solar strategy.

Lighting the Political Touchpaper

The launch of mass and series production is urgent because it represents the quickest way of bringing about the shift towards a solar economy that might be supported by business and society, without having to wait for political initiatives at each step of the way. The job of these institutions should be limited to providing effective launch motivation and to establishing a general frame of reference. A distinction must be made between direct and indirect, cost-effective, zero-cost and even cost-reducing initiatives for the market introduction of solar energy. To the extent that they are cost-effective, the suggestions should be limited to those that can be financed by a change of priorities and thus are cost-neutral in the final outcome.

Political Institutions Should Set the Example

Political institutions use large amounts of energy, mainly for heating and air conditioning of public buildings, for operating government car and utility vehicle pools, and in providing and operating public infrastructure. There is no reason why this sector should not immediately begin to meet this energy demand through solar energy. Political institutions should be bound by law to equip all new or about-to-be-renovated public buildings with solar roofs and solar facades – collectors and/or solar cells – and to meet design standards for 'minimum emission buildings', including the use of solar energy. The costs would be recovered over time by savings from not using conventional facade elements, and by energy savings. The initial extra financial requirements for solar cells will have to be borne because the public sector must set an example, quite apart from the fact that these costs could be offset by savings in other parts of the total building costs. After all, nobody questions whether the extra outlay for this lobby or that stairway system or that particular construction material is 'competitive'.

It would send a strong signal if prestigious major public buildings could be equipped with solar facilities, setting a course for the future. The newly refurbished Reichstag in Berlin, housing the German Parliament, has done this. However, the new buildings for the European Parliament in Brussels and Strasbourg are the same old energy guzzlers as most of the buildings of the last few decades. Such structures attract a great deal of public interest. If the home of the highest democratic constitutional authority is a showcase for solar energy, it becomes living evidence of a political programme for the future, and is likely to stimulate countless private efforts that otherwise could be launched only with the help of expensive public subsidies. In addition to the public effect of setting such an example – and the gain for the environment – the following effects are likely:

- It would immediately create a market for solar collectors, cells and other solar building elements. Demand for these technologies would increase, and the cost of solar technology equipment for the general market would drop.

- The construction trades would begin to take these new technologies into account, enabling them to handle new jobs for the private sector in this field.

- Other building owners would follow these trends.

- Architects would increasingly employ their skills to create similar buildings.

The specifications for new public project awards must state that all newly installed, free-standing public illumination equipment – street lights, traffic signs, phone booths, bus and rail stops and others – are to be operated exclusively with solar cells, saving cabling costs. Examples of this kind already exist, but there is no general mandate for their installation that could lead to a broad-based effort and jump-start production. New vehicles in government motor pools used for short-range transport should be electric vehicles equipped with appropriate power generation facilities for fuels from solar or wind sources.

Plunging into Mass Production of Photovoltaics

Above all, in order to set in motion the production of solar cells, which desperately needs market stimulation if they are to be manufactured with greater cost efficiency, a government procurement programme of some years' duration should be organized for a country's solar cell producers. The first formulated mass introduction programme was the EUROSOLAR proposal in 1994 for 100,000 rooftops and facades, which started in 1999, after the German federal election in 1998. Through this programme, 100,000 photovoltaic rooftop and facade installations are to be subsidized by a state financial incentive for private sector contractors, in order to encourage the leap into mass production. Together with the new German feed-in law, the German Renewable Energy Act, it will be possible to drive down the production costs for PV and strengthen mass production.

In 1996, Japan started a similar large-scale programme to install about 70,000 rooftops. In 1997, President Clinton announced a one-million rooftop programme, although no resources were allocated to it in the federal budget. It relies on the motivation of states, cities and the private sector to reach the target quota. EUROSOLAR's demand for a 1 million installation programme for the European Union, which was incorporated in its advice to the EU White Paper on renewable energies, was taken up by the European Union parliament in its recommendations on the subject in May 1997 (in the report by the member of the European Parliament (MEP), Mechthild Rothe, the European Parliament took the vote. At the end of 1997 the 1 million rooftop programme was indeed recommended by the White Paper, which had been taken over by the EU commission. However, so far no funding decision has been taken.

Zero-Cost Public Measures

Measures that merely require political will to create something new, but no public expenditure, should initially consist of efforts to remove the administrative obstacles that stand in the way of solar energy. These include, above all, regulations and laws covering construction, which should require:

- a legal obligation for planning permission to demand that new houses and groups of buildings have their main aspect facing south, and that the building's outline be oriented according to the solar radiation's angle of incidence;

- a requirement for an energy rating based on the experience with 'low-emission buildings', which would provide binding target criteria;

- abolition of licences or permits for the installation of solar facilities in all types of buildings covered by all building regulations, and for stand-alone wind power facilities in agricultural areas; priority to be given to building permits for wind power facilities, biomass installations and running-water hydro plants without time limits;

- a clear description in nature protection laws of facilities using renewable energies as essential elements of nature preservation, so that renewable energies become a positive factor to be considered;

- a change in budget rules covering public buildings' energy facilities, to base them not on current cost comparisons, but on a ten-year cost–benefit analysis.

Beyond that, a regulation would be appropriate that would prescribe for every new building and for any re-roofing the installation of a minimum capacity of solar collectors or photovoltaic facilities. In the case of solar thermal installations this would not be an unreasonable burden on owners even today, especially since this criterion of 'reasonableness' is not a factor in other, less important construction codes. The large new market that would spring up overnight because of this would quickly equalize the present slight cost differences from those of conventional heat and electricity supply systems. Why should building codes covering the visual aspects of a cityscape have a higher standard of what is reasonable than those that address clean air and climate preservation?

All this would also ensure that solar energy use in conjunction with energy-saving measures would automatically become part and parcel of overall construction planning by owners, architects, construction firms, construction planning and zoning agencies, local authorities and residents – in other words, in its totality it would support a positive attitude on the part of building professionals to natural living conditions. Solar energy use in and on buildings must be as routine in the construction process as the inclusion of windows, doors and plumbing. The ancient cultures, from Egypt to pre-Colombian America, have already shown how to do that. We cannot claim to be civilized nations if we don't follow these traditions with modern, and much easier to use, technical means.

Another cost-free measure would be to allow only solar-powered or biofuel-powered private motor vessels on rivers, canals, lakes and offshore waters. This would be less radical than existing outright bans on motor boats powered by fossil fuel in force on some inland waterways. Again, this would almost automatically create a new market. The measure would affect only those who pollute public waters and the air at the expense of the general public. A programme to replace motor boats by solar boats should be started perhaps for the city of Venice, whose unique and threatened buildings are being further damaged by power boat vibrations: this would be a signal that would attract the attention of the rest of the world. A study for EUROSOLAR in 1995 demonstrated this possibility for Venice. This city was chosen in order to help attract broad international attention to solar-powered boats.[167]

Permission to reactivate and construct small hydro power plants should be granted more generously. New regulations should be promulgated to mandate the use of gases generated by all rubbish dumps and the energy utilization of organic wastes in sewage plants, waste removal facilities, and large agricultural operations. Again, this would turn the utilization of these energy potentials into everyday issues of every planning programme. Mandatory obligations create less of an administrative effort than a multitude of varying individual rules and solutions.

There must be a way of overcoming the present unhappy situation, which relies on haphazard individual initiatives, in order to make common practice what is both possible and already practically demonstrated today.

Transforming the Energy Industry

In the United States in particular, utilities have been switching to 'least-cost planning', which is being practised with increasing success if only on a voluntary basis.[168] This approach was stimulated by the recognition that it is cheaper for utilities to finance energy-saving efforts by consumers than to spend more on new peak-load capacities – plants that generate energy only at times of very high demand and are therefore unprofitable. California has developed a system of competitive tendering for the supply of peak-load energy. The supplier who can meet that demand at least cost, for example by using previously unexploited heat sources and by shifting to more cost-efficient power generation from natural gas, wins the contract. This financing concept of 'negawatt' instead of 'megawatt' can achieve considerable energy savings in a relatively short period of time. The goal of 'negawatts' of environmentally destructive conventional energy obviously cannot be achieved solely in this way, because even the most efficient conventional plant needs fossil fuels, and the thriftiest consumer appliance needs a minimum of electric power. 'Least-cost planning' therefore must be augmented by other, new structural principles within the electricity supply industry to launch the transition to solar energy plants.

Statutory Guarantees for Feeding Solar Power into the Grid

The right to feed privately generated electricity from solar power plants into the public grid must be guaranteed by law at a price that corresponds to the average power generation costs of utilities. This is a step that has been taken in Germany, for example, through the Electricity Feed Law for Renewable Energies. This law has been in force since 1 January 1991. In 2000 this law was replaced by the Renewable Energy Act with much better price conditions. For electricity from photovoltaic devices the payment is 99 German Pfennig/kWh, from wind it is 17.8 Pfennig, from Biomass between 17 and 20 Pfennig, from small hydro plants 15 Pfennig.

On the basis of the German Electricity Feed Law, 6,000 MW of wind power facilities had been installed by the end of 2000. This proves how quickly renewable energies can be developed by private operators. In turn, this increases the pressure on the EU for a directive for electricity feed from renewable energies. In fact, such a directive was recommended by the EU Commission in its White Paper on renewable energies in 1997, on the basis of a similar idea from EUROSOLAR's 1996 study about the electricity feed regulations of the EU countries.[169] The proposed compensation criterion is that a distribution firm should pay at least the average draw-down price of conventional electricity plus an environmental premium of 20%.

This could revolutionize the electricity production sector, and ensure a rapid transition from central large-scale power stations to innumerable small power plants.

Community Energy Supply

In general, the goal is of local and municipal energy self-government. This will bring about a revitalization of municipally owned or local utilities, which have largely been reduced to the role of dependent distributors – if they still exist. Where such municipal utilities no longer exist, new ones must be established. The repurchase of electricity networks for municipal distribution is the decisive step in this process. The production of electricity and heating, inasmuch as they are not provided by solar thermal facilities, does not need to be under the control of the municipal utilities at one and the same time. It would make sense to have a municipal distributor and a municipal generator unless electricity production was available from private operators at the local or regional level.

Since solar energy utilization from biomass consists in large part of by-products from other functions, it offers a wide spectrum of tasks for these municipal or local utilities: the integration of electric power, gas and heat production and waste and sewage removal.

One job of such a utility could be, for example, to obtain user rights for the facades and roofs of existing buildings, and to install and operate solar technology equipment if the owners were reluctant to do so. Communities could also assume the job – for the time being at least – of converting biomass into biofuels or biogas by contracting the work out to private companies, by launching such companies themselves, or by assisting local producers in marketing these products. Municipally operated utilities are in an ideal position to generate electricity and heat from biomass in cogeneration plants.

These utilities would be able to organize a careful transition to a solar energy supply system by operating initially a combination of solar energy and conventional energy. At the same time, they would stimulate local and regional industry because the know-how needed for these technologies is easily acquired. Such a communal approach would also be easy on the taxpayer since it would not require any big leaps in investment. With this strategy, new energy components could be added, one at a time, to the existing ones.[170] Figure 5, on page 192 shows the Austrian Federal State of Steiermark as an example of how such a concept evolves into a widely spread, decentralized energy system.

Inter-regional generating companies offering energy through the grid would provide local utilities with a progressively smaller load-balancing and back-up function as local solar energy utilization expanded. Nevertheless, they would continue to fill an important role because trans-regional load balancing of solar-generated power will be an important component of a solar energy economy. The precondition for

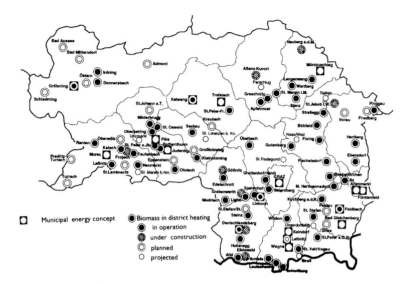

Figure 5 *Biomass district-heating networks in the Steiermark region (Austria)*

Source: Landesenergieverein (State Energy Association), Graz. Graphics: J. Srienc

such a reorientation towards municipal and local energy utilities would be that they be prohibited by law from signing long-term delivery contracts with trans-regional energy supply companies, which would obstruct the otherwise continuously growing production of final energy from solar power. Contractual obligations must not prevent these local companies from substituting conventional energy supplies as quickly as possible.

Freeze on New Conventional Power Plants

What would be called for is an investment freeze on the construction of new nuclear-, coal- and oil-powered generating facilities by the simple expedient of no longer issuing construction permits for these facilities. In other words, fossil or nuclear power plants, which do not use the excess heat from electricity generation with steam turbines, are and will remain *per se* extremely ineffective, and should no longer be given permits in the future. In this way, nuclear facilities would be out of the question, and fossil fuel plants could be permitted only in small measure as cogeneration plants. Guiding investments in this fashion would mean that:

- no longer would huge investment funds be generated for nuclear-, coal- and oil-fired power plants that would produce additional decades of operations of such plants to assure the return flow of capital;

- efforts to invest in energy conservation would receive a big boost;

- solar energies would account for an increasingly large slice of energy investment. For the construction of new capacity, utilities might already be inclined to tackle the potentials of wind, solar cell and solar thermal power generation as well as biomass utilization.

Solar Structural Change Rescuing Public Finances

Virtually every national budget is suffering from the burdens of permanent public subsidies. This commitment of public funds ties up disproportionately large amounts of money at the expense of planning for the future. This situation is perpetuated either by claiming material social needs for those affected, or with immaterial arguments of national prestige, technological competitiveness or security against threats from abroad. Four key examples will demonstrate how these subsidies can be reduced in order to benefit both the common good and the individuals concerned, and, at the same time, how it can be linked to the introduction of solar energy.

Agriculture as Standard Bearer of Solar Energy

As is widely known, the European Union's agricultural subsidies have reached outrageous proportions without achieving the goal of efficiently and successfully sustaining agricultural structures. A programme is both possible and overdue that would enable agricultural enterprises to install biogas plants with generous subsidies, or make possible the construction of plants for the exploitation of other agricultural or forest product waste, or even for wind or hydro power facilities. Such subsidies could take the form of interest-free government pre-financing, offset against expected future agricultural subsidies. The advantage for farmers would be a long-term, stable existence guaranteed by savings from redundant outside energy purchases and by income from the sale of energy themselves, while continuing with agricultural production. After all, anybody who uses agricultural wastes is more likely to stay in the agricultural business. Instead of forcing farmers into the role of paid conservators of the land, they would be turned into operators of environmentally friendly energy supply facilities – economic guardians of nature's cycle. As they are gaining an additional, stable source of income, prevailing agricultural subsidies should be gradually eliminated over the medium term.

Such a programme would have a multiple effect: reduction of CO_2 emissions into the atmosphere; prevention of considerable emissions of methane into the atmosphere, since it would be burned to generate energy; protection of the soil and of groundwater against excessive fertilization from liquid manure; a contribution to waste and sewage sludge removal; production of high-value, environmentally friendly fertilizer and plant protection from fermented liquid manure; an increase in agricultural income and assistance in safeguarding the existence of farmers and, with it, the opportunity to reduce agricultural subsidies in the future; and the creation of a new production branch in plant construction. Almost any one of these effects is, in itself, a valid reason to follow the course of biogas utilization – instead of continuing to meet energy demand via oil delivery and combustion, of permitting methane to aid in the destruction of the atmosphere, of devastating soil and groundwater, or of channelling manure and sludge into rivers and lakes and subsidizing the endangered existence of agriculture with public funds.

For cities in developing countries, which frequently do not have a reasonable means to dispose of organic waste, there are other advantages. Using these wastes to generate energy would make a crucial contribution to public hygiene and thus help fight the spread of epidemics. Cities such as Mexico City suffer the stench and the dangers of infections from the organic waste of over 20 million people. Instead of thinking of this waste as an energy resource, and thus significantly increasing the quality of life, local planners still think more in terms of constructing nuclear power plants for their municipal energy supply while there is allegedly no money available for the removal of organic wastes.

This opportunity to reduce subsidies becomes even more pronounced with the cultivation of energy crops: instead of farmers being paid to let their land lie fallow, they would receive a purchase guarantee – just as for cereals or milk – for the cultivation of selected types of high-yield energy crops that require little or no fertilizer or pesticides. This would enable them to plant fast-growth forests and reed grass, because they would have access to energy markets. The result would be immediate relief for agricultural budgets because, in the course of this restructuring operation, public outlays for storage and destruction of agricultural products would be reduced.

Such trends should produce new agricultural structures that lead to economic self-sufficiency and a reduction of subsidies. This follows from a rough 'input–output' calculation. Approximately two thirds of total sales of an agricultural enterprise are spent on inputs such as feed, fertilizer, insecticide and herbicide sprays, machinery and fuel. Biogas plants would greatly reduce outlays on fertilizers and would entirely eliminate fuel costs, which account for as much as 20% of overall operating costs. On the other hand, revenues would increase from processing organic wastes for other farmers and from the sale of electricity. If energy crops are cultivated as well, the costs for fertilizer and sprays will decrease further, and the income increase. As soon as new production and marketing cooperatives for bio-energy have been set up and energy cooperation has been established with municipal utilities, new regional structures that could be integrated into industrial society could emerge, along with a durable, stable agricultural identity. The bleeding of rural areas can come to an end. Table 19 on page 196 shows the social, environmental and economic advantages to be derived from the economic utilization of an energy forest.

A financial starting point for the introduction of such conversion plants could be established by converting existing energy subsidies for agriculture into an effective investment stimulus. In Germany, for instance, about $625 million are paid out annually in subsidies to make gas oil cheaper. These subsidies could be eliminated and replaced by interest-free

Table 19 *Social, environmental and economic advantages of biomass utilization (based on the example of an energy coppice)*

For individuals

Per 500 tonnes of fuel sold
 – 1 full-time harvest job
 – 1 full-time forestry job
 – 1 full-time administrative job

For the environment

Locally
 – improvement to the natural habitat
 – improved viability of agriculture
Nationally/globally
 – reduction of CO_2 emissions
 – reduction of SO_2 emissions

For the local community

Halt to depopulation
Money spent on energy remains in the local economy
Additional users for public and private local services
Employment prospects for village youth
Continuity between generations for local families

For public finances

Savings on
 – social security benefits
 – family support
 – housing benefit
 – increased tax revenue through employment

Source: Clare T. Lukehurst, 'The economic significance of using biomass', in *Feedstocks for Energy Production in Rural Areas, Proceedings of the International Conference on Biomass for Energy and Industry*, Florence, Italy, 5–9 October 1992

loans for investing in biomass facilities. Assuming an interest rate of 10%, only the interest would have to be paid out of public funds; the loan itself would be paid back in annual increments over 10 years. This approach would make available ten times the volume of interest subsidized. Farmers taking advantage of this scheme would invest in the future, and have available enough self-generated energy to meet their own needs and an additional source of income. Those who did not take advantage of this offer would be disadvantaged by the elimination of the gas oil subsidy.

From Weapons Technology to Solar Technology

The development and production of weapons technologies in NATO member states and Russia has barely been interrupted, despite the end of the East–West conflict. Neither governments nor defence industries know what they could do otherwise. Because of this lack of imagination, billions and billions more of public funds are being squandered. In order to legitimize the waste, the West has invented new dangers, in order to be able to introduce new weapons. The new trend is towards rapid force deployment, for which *inter alia* new deployment systems, such as fully equipped helicopters and transport aircraft, are required. Missile defence systems also continue to be developed.

The structural difference between the defence industry and other industry segments is that the industry's military products have only one customer: government. No business sector, therefore, is easier to redirect politically into other areas of production by replacing military orders with civilian ones – in other words, by changing the priorities from external security to environmental safety. Obviously it is not a problem to build civil aircraft powered by environmentally friendly fuel instead of fighter planes. Government as source and sponsor of mega-projects can direct cooperation among industries in these areas as well, just as they always have done in the procurement of military hardware.

For reasons of structural policy, the political task now is to channel civilian technology projects appropriate to its technological know-how into the defence industry instead of further defence-related projects. Here, solar energy offers a number of opportunities. As alternative technologies would have to be sold on the private market, in contrast to weapons programmes, this would reduce the pressure on national budgets within the next 10 years. Some examples are:

- in the area of shipbuilding, the design of ships with bio-fuel propulsion systems, new freighters with electronically controlled sails, floating solar cell pontoons, or solar desalination plants;

- in aviation, the development of hydrogen-powered aircraft (the first test flight of such an aircraft occurred in 1988 in Moscow); and of new dirigibles that could serve as efficient transport craft;
- the development of new rotors for wind power plants;
- the development of lightweight materials for aircraft or railway rolling stock, producing energy savings.

From the Coal Penny to the Solar Penny

Some $12.5 billion is being spent annually in the European Union in subsidizing coal production in Germany, Great Britain and Spain, none of which can compete any longer with coal prices on the world market. The official reason is for the security of domestic energy supply, something that can be achieved to a much greater extent with solar energy. The real reason is to safeguard existing jobs, although the subsidy costs for these jobs no longer bear any defensible relationship to the results. Some $43.75 billion in coal subsidies will have been spent up to 2005 to maintain about 50,000 jobs in German coal mining operations. This annual expenditure is about forty times the amount in public funds that will be spent on solar technology R&D. This is no longer defensible for an energy source that is among those that need to be replaced. Even if full allowance is made for the goal of preserving jobs, it should be possible to overcome this contradiction.

One possibility would be step-by-step reduction of coal subsidies, and the use of funds released over several years for investment aid to help construct production facilities and to retrain personnel in the coal mining areas for work in solar energy production facilities and energy savings technologies. At the same time, such investment aid would represent a technological prioritization programme on an unprecedented scale for solar technology, and thus an accelerated timetable for its introduction. It would mean that in spite of the expiry of coal subsidies no job losses would ensue. On the contrary, alternative jobs would be created with considerably greater job security than in mining with coal subsidies. They would be jobs with a future – and much healthier jobs.[171]

This investment aid should be available to all interested investors, but in the initial phases it should be tied to initiatives in coal mining regions. As soon as a sufficient number of replacement jobs had been created, the programme could either be terminated ahead of time, producing even more savings in subsidies, or it could be expanded to all regions within the European Union. This would represent a model of active structural policies for safeguarding both employment and nature, as well as for the recovery of government budgets. Last but not least, the result would be more in line with free market economics.

Financing the Solar Energy Offensive

In order to circumvent the reaction of the major financial institutions against the danger of capital annihilation by solar competition that (from their perspective) is appearing on the scene too early, the questions of where the willing investors are or what might be suitable financing pathways for introducing solar energy must be answered. Naturally it would be astute strategy for the future if corporations were to diversify as soon as possible. If a major car-maker were to diversify into solar-powered electric cars and cogeneration plant manufacture, it would probably have a much brighter future in an infinitely large market than, say, Daimler-Benz, which diversified into defence and space technology and thereby added further to the old burdens of unsustainable structures. Everybody can buy solar technology, but not rockets, fighter planes and satellites. Thus the investors in a radical solar economy remedy, in terms of equipment manufacture, are most likely to be those who have been least integrated into the traditional energy structure: electrical equipment manufacturers without a stake in nuclear plant construction; monostructural (non-diversified) car manufacturers; and glass and ceramics manufacturers as well as the construction materials industry because of their familiarity with existing market structures.

Who would finance a solar energy offensive on behalf of potential investors without large capital resources of their own? In this regard, new financial institutions have been established in various countries in recent years that offer relatively low-interest environmental credits. But these are still small institutions, without the clout for a major offensive. Other banks are beginning to offer 'green capital' at 1–2% below prevailing interest rates. But the amount of capital that must be mobilized for a solar offensive is so large that these beginnings cannot hope to meet the need.

A major step would be to tap the assets of the insurance companies for the solar structural change. Part of the insurance business, especially the life insurance business, is the formation of long-term reserves – and only investments in those corporations that manufacture environmental products are likely to be profitable in the long term. Large investment groups are beginning to recognize that unaffordable risks are coming their way with the continuation of the current energy system. Since insurance companies can least afford to give preference to quick cash over long-term returns on investment, every effort should be made to win them over to the idea of structural changes in the energy business. They should invest a portion of the capital generated from their insurance customers in companies that produce or operate solar energy technology. The Swiss financial expert Kurt Müller sees this as a strategic opening.[172] In the past few years, some insurance companies, especially those involved in the re-insurance business, have been giving some thought to these basic considerations. The fact that it has not yet been grasped by all insurers, and life insurers in particular, is probably associated with their intensely close relationship with the major banks.

The concept of solar banks and solar credits, as suggested by EUROSOLAR, would be suitable for financing the investment in solar energy facilities, both for individuals and for utilities.[173] These credits would have to take into account the specific characteristics of a solar facility: that is, while there is a need for capital to finance the facility itself, there would be no energy operating costs – except for the cultivation of energy crops. Thus the saved energy costs could be set off against loan repayment, which in turn means that the operator could afford to take on a larger investment burden compared with conventional energy facilities, without any additional financial strain. The combination of such long-term credit financing with consumer solar energy advice would almost immediately create a solar technology boom, starting with a switch to solar energy by homeowners and the owners of commercial buildings. Just as cooperative home-building societies and systems of home construction credits have been created in the past to meet the basic human need for shelter, the same thing must happen to meet the basic need for clean energy. Nor is it absolutely necessary to set up new banking institutions: specialized departments in banks and building societies could do the job. It would be part of the solar strategy to reshape national, regional and local banking institutions along those lines. The widespread network of independent cooperative banks could serve this purpose as well.

Independent Initiatives

To wait for parliaments, governments and communities to enact the required key legislation is unacceptable. It would make solar strategy dependent on political institutions, and that would not be enough, especially in the current historical context, where these institutions are finding it increasingly difficult to adapt to the real problems. To move a solar strategy forward, an active society is essential; it needs independent initiatives to break through constraints that allegedly exist, and to create projects. Such a strategy, which supposedly leads to a decentralization of structures, needs impetus from committed individuals who not only demand these alternatives but who will themselves turn them into reality.

It is no accident that the solar showpieces that already exist are usually the result of such individual efforts: the solar thermal power plants in California, wind power plants, biogas plants in small agricultural operations, and solar homes. Also, advice to citizens whose interest in solar technology applications has not been satisfied by commercial or public technical advisers has been provided mainly through public interest initiatives. Historically, this is not new; the social movements of the 19th century had similar experiences under much more difficult conditions.

Independent cooperatives played major roles then in areas such as purchasing, production, marketing, credit, housing construction and consumer cooperatives.[174] This movement was very successful, and in many cases grew too large, became bureaucratized and lost its idealist motivation. The movement positively influenced societal structures but in the process integrated itself into society so thoroughly that its organization was hardly different from that of any other business. New, independent initiatives must be shaped for the new goals and interests that lie ahead – those that have not been addressed at all or marginally, at best, by established political and economic forces despite the fact that there is demand for efforts of this kind.

The use of solar energy is the most appropriate example imaginable for this kind of activity. Operators' cooperatives are a model of how a carefully targeted revitalization of cooperative initiatives can be achieved in the agricultural sector. Other examples are the formation of joint stock companies for renewable energy investments, or the do-it-yourself movement for constructing solar collectors, both of which are part of the spectrum of such initiatives. Because such projects involve a relatively large number of small investments in energy facilities and not just a few large investments, almost anybody can participate in their formation. And since everybody can actively take part, even on an individual basis, a solar strategy is 'open' in terms of public involvement – more so than ever before with any comparable issue affecting the prospects for the future. It is possible to evolve a new culture, in which mankind is not merely one cog functioning in a mega-machine. And if a kind of mutually accelerating interaction should evolve via the cultural revolution 'from below' and a revolution of political decision-making processes 'from above', the implementation of such a solar strategy will become unstoppable, current obstacles notwithstanding. It will become possible to undermine the traditional energy system with highly efficient small-technology systems, and to launch a rebellion with thousands of individual steps that will evolve into a revolution of millions of individual steps. This can be achieved with low-tech cogeneration systems, from individual households to businesses and trade operations – for example, by using biomass combustible material.

With these practical and very realistic efforts – and with the help of new companies developing and manufacturing these technologies outside the energy cartel – a solar strategy can be launched even if the political structures remain too inflexible.

Its realization will take somewhat longer, however, because the first batch of equipment – before the start of mass production – will be disproportionately expensive. It will enter the market only via those customers who have consciously altered their private priorities. Despite this opportunity for a practical citizens' movement, as formulated here, radical political change is still needed. There is no more time to be lost. With independent initiatives, only a few of which are mentioned here, humanity has an opportunity to cease to be the victim of the catastrophic consequences of nuclear and fossil energy provision and to become the instigator of the solution to the problem. The more this is the case, the more infectious it becomes and the wider the civic movement will be.

8

The Entropy Tax

The most universal measure to promote energy conservation and the transition to solar energy across the entire spectrum of energy consumption is undoubtedly higher taxation of traditional energy. Such a legislative step, designed to compensate for the social costs of this energy, is instrumentally the simplest measure to implement. However, at the same time it requires the greatest courage and foresight, if it is to be comprehensive and therefore sufficiently effective. Just taking the decision will actually be extremely difficult, as it will meet with the greatest and most diverse opposition.

The key political factors for the introduction of solar energy discussed in the previous chapter do not depend on higher energy taxes. They contradict the widespread prejudice that higher taxes on traditional energy are a necessary precondition for the introduction of solar energy – a view that provides many with a convenient excuse for doing nothing.

A huge opportunity was missed in the 1980s, when oil prices dropped again after the collapse of the OPEC cartel. At the time, hardly any countries attempted to compensate for these price falls with higher taxes on crude oil, although many people had already become used to higher energy prices. The only positive exception was Denmark. People preferred to worship the Golden Calf again rather than attempt to reform energy demand with the help of fiscal policy. The energy tax debate was revived at the end of the 1980s because of the early signs of a coming climate catastrophe. Previously, energy taxes had been designed primarily as cash cows for government coffers – though always linked to the goal of making energy available to industry as cheaply as possible. This in turn dictated the method of taxation and the differences in energy pricing: the basic idea was to keep energy prices as low as possible for industrial production, but relatively high for the final consumer – for example, car users and domestic electricity customers. Most of the suggestions on how to save and tax energy focused on the final individual consumer. The new aspect of environmentally motivated energy taxes is to do with the basic reform of the energy system, and not with a desire to find more income for government. Higher energy taxes should therefore go hand in hand with a reduction in other taxes or expenditures, and the environmental tax criterion will lead to a change in the entire tax system.

Such a change can take place progressively, for example by first dismantling the countless fiscal privileges enjoyed by nuclear and/or fossil energy utilization. Alternatively, a system of targeted taxation can be created over time, focusing on individual energy sources and differentiating between them according to their production of harmful substances, or other specific criteria. However, there is a danger of systematic dilution of energy taxation concepts. This dilution is a regular occurrence. In most energy taxation systems based on ecological concerns, industry is provisionally exempt. The official reason for this is the desire not to compromise its international competitive position. In fact such exceptions result from the power of particularly well-organized interest groups. After all, the potential for rationalizing energy use is just as under-exploited in industry as it is in the public sector or in private households.

There is a great deal of latitude for this everywhere, as shown by the difference in petrol and diesel fuel taxes in different countries. Italy's fuel taxes are almost eight times higher than those in the United States; but the lively pace of car traffic in Italy demonstrates that higher taxes do not necessarily lead to less driving. Instead, buyers switch to smaller cars – since other possible changes, such as to solar cars for instance, are still in their infancy. In areas where an immediate switch to solar energy is possible – domestic heating, for instance – taxes on oil and gas are clearly lower than those on vehicle fuels. Therefore, in order to achieve an effective fiscal stimulus for solar energy, energy prices would have to be raised significantly across the board.[175]

In 1992, the European Union wanted to introduce a CO_2 tax, supplemented by a tax on nuclear power, which would otherwise have a competitive advantage. That effort has failed for the time being because of strong internal opposition: initially, member countries such as Spain and Portugal blocked the new taxes, believing they needed to catch up with the other member countries in terms of environmentally damaging economic growth, while another argument was that a CO_2 tax would be viable only if the United States and Japan also had one. The argument that low energy prices are essential for international competitiveness keeps occurring, although it is a simplification without merit: the United States clearly enjoys lower energy prices than Germany and Japan without damaging the competitive position of German or Japanese products.[176]

Environmental tax reforms usually aim to reduce drastically taxes on wages and incomes as offsetting factors – to compensate higher tax burdens with relief elsewhere. Farel Bradbury and Malcolm Slesser go the furthest of all with their proposal for a 'Unitax': they want to replace wage and income taxes, all corporate taxes, consumption taxes and social levies with an energy tax placed on primary energy at the point where it enters a nation's economic cycle, and calculated on the basis of past total revenues collected by all public treasuries.[177] But a total overhaul of the tax system such as this would require one undifferentiated energy tax that would have to include solar energy, otherwise the system would not survive because, following the wholesale conversion to solar energy, tax revenues would decline too far, and after a while the entire tax system would have to be totally revised once more. The impressive feature of the Unitax idea is the radical tax simplification; as energy taxes would be collected at the start of the energy chain, in a national economy the internal revenue service and other national tax collection agencies would shrink to mini-agencies, and the bureaucratic extravagance of having companies and individuals fill in income tax forms would disappear.

On the other hand, misgivings have been voiced about offsetting a higher energy tax with lower income taxes, as that approach would unfairly create additional burdens for lower income groups. Those who pay no income tax, such as the unemployed and retired, would merely suffer higher energy costs, without the corresponding benefit of any tax relief. As a concrete example for the advantages of Unitax, Slesser cites the example of a British family with an annual income of £20,000; it currently pays about £6,400 in income taxes and social security contributions. Under Unitax this family would pay instead an extra £1,400 for driving a family car and £3,500 for heating. As other value-added taxes would also disappear, the family would in effect enjoy tax relief. This example does not address the twin drawbacks of a significant extra tax burden on non-taxpaying citizens, and drastic tax reductions for high-income earners. Possible adjustments for the less well off, such as heating subsidies, would again require considerable administrative effort.

The New Principle: Taxation of Non-Renewable Resources

The central questions of an environmentally based tax system are:

- What is the generally applicable principle behind the concept for a new tax system?

- What other taxes should be reduced?

- At which point in the economic cycle should the tax be collected?

The basic criteria are:

- environmentally oriented economic dynamics;

- energy and product prices that correspond more closely to social and environmental reality;

- social justice in energy taxation;

- avoidance of further complication of the tax laws and – if possible – simplification.

Enforcing an environmental concept of taxation depends mainly on whether it is simple, and free of any inherent contradictions. Environmental taxes must not be dissipated by countless individual tax regulations that would make the entire system impenetrable and produce an impracticable mountain of tax forms and unworkable tax assessments.

These suggestions, taking into account the basic principle of safeguarding the environmental and economic future (reducing the level of entropy created by human activity) plus the resulting Energy Imperative (see Chapter 2), lead to the basic concept of taxing entropy. The resulting measures should be implemented at the European Union level; it urgently needs to adjust the various national tax systems to a common denominator, something that will be simpler with new tax principles, rather than by assimilating traditional tax systems that are not really comparable. It is precisely in the area of tax policy that the European Union faces chaos if this is not achieved – with innumerable possibilities for tax evasion, especially by multinational corporations.

An entropy tax could provide systematic control of all environmental damage, and would permit its social importance to be determined. Since all aspects of taxation would be linked to the extraction, production and use of energy or other goods, it is important for the sake of supporting environmentally based economic dynamics and social justice to offer relief on other consumption and corporate taxes. Thus an entropy tax would systematically cover all the damage done by non-renewable resources. The plan is thus:

- to start initially by taxing conventional energy and, in turn, giving up value-added tax; alternatively, to introduce a very high level of value-added tax on energy at first and, in return, reduce the rate for other products;

- as the second, third and fourth steps, to tax mineral raw materials, environmentally damaging chemical materials and the use of land, removing corporate taxes in return.

Taxation should occur when the factor that is to be taxed enters the economic cycle: that is, where customs or tax jurisdiction, or the commercialization of land, sets in. Only in this way can there be an immediate effect by such a tax on those production processes whose modification would be the pivotal point of an environmental economic dynamic. It would also greatly simplify the calculation of taxes, not only for the tax administration but also for business especially, if the compensatory tax breaks suggested were adopted.

The New Economic Dynamics

This concept could be realized in the following steps, and would have the following effects.

Stage 1: Taxing Traditional Energies

Oil would be taxed at the refineries, which always process both imported and domestic oil; coal would be taxed at the importer and at the domestic mining companies; natural gas also at the importer or the domestic producer; nuclear power taxes would be imposed at the level of the uranium enrichment operators or at the level of the nuclear plant operators. These drastically increased prices for primary energy would automatically be passed on to industrial enterprises, electricity producers and oil companies, but would still permit differentiation between heating oil and vehicle fuels.

In general, a massive push towards more efficient energy use and energy conservation, and a switch to tax-free solar energy, would get under way – and it would also relieve pressure on the balance of payments. Eliminating value-added taxes as a compensatory move would produce several major advantages: energy-consuming companies would get administrative relief, and the focus of their business activities would shift to reduce energy costs. Prices would increase only in those instances where additional energy costs were higher than the previous value-added tax rates; in other cases they would drop. The same would hold true for private final consumers of energy and other products. Environmentally damaging high energy use would always be 'punished', while low energy consumption would bring marketing advantages. The additional burdens and the new relief for both industrial and private consumers would be roughly balanced, modifying the choice of products. The administrative load on government would be reduced.

This approach differs from other frequently discussed energy taxes as industrial energy consumption would be fully covered and finally as value-added taxes would be reduced or abolished in return. The total value of the extra energy taxes thus would be equated with the revenue from value-added taxes. The amounts of revenue involved can be demonstrated by the example of German revenue streams derived from value-added tax. In 1996, the total came to $125.2 billion; in contrast, the mineral oil tax, including taxes on heating oil and natural gas, totalled $42.7 billion. Elimination of value-added tax in favour of a higher energy tax would mean almost quadrupling the energy tax to

$167.5 billion. The revenue would almost immediately drop again because of extra energy savings and a switch to renewable energy. The resulting shortfall would be compensated for by the impetus given to new products and, with them, new jobs, followed by higher tax revenues, lower welfare costs and a reduction of environmental costs. Furthermore it would not contradict the concept if value-added tax for other products and services were to rise again after initially being lowered, as soon as the conversion effect had been achieved and secured.

Stages 2, 3 and 4: Taxation of Other Entropy Factors

Choosing which of the next steps should take place first involves a careful weighing of further alternatives and appropriate systematic preparations, given the multitude of mineral raw materials and chemicals. A variety of approaches will be required.

The main consumers of industrial raw materials and also the main importers – mostly from developing countries – are Japan, Europe and the United States. Prices are one of the central problems in the ferocious economic war between North and South. Developing countries depend on revenues from their raw materials but, for all practical purposes, they are victims of the low-price dictatorship of the buyers who exploit their industrial monopoly. This leads to continuously rising political tensions.[178] The rapid exploitation of raw material deposits runs counter to the long-term interests of both developing and industrial countries. Since those minerals cannot be plucked from the earth in their pure state, the extraction of usable raw materials requires considerable energy input, and it leaves other environmental damage in its wake, mainly in developing countries. It is certainly likely that the energy balance of raw material extraction and preparation in developing countries is negative; in other words, the local economies suffer more damage from the energy imports required for these purposes, and the consequent forest destruction, than they benefit from their exports. As long as raw materials are cheap, the opportunities for recycling are not exploited, although the initial use of these materials generally requires more energy than their subsequent reuse.

Because of low raw material prices, this secondary use of materials decreased between 1950 and 1980. In the United States it dropped for aluminium from 51.4% (1950) to merely 41.1%; in Great Britain from 18.9% to 4.7%; in Germany from 13.0% to 11.3%; and in France from 5.5% to 4.4%. Only in Japan did it increase, from 0.8% to 19.9%. Similar trends were noted for copper and zinc.[179] Energy demand and waste have played virtually no role in energy balances and environmental analyses until now,[180] although they represent a substantial component of the entropy increase. The metal industry alone contributes about 10% of global CO_2 emissions, and in Brazil, for example, stretches of jungle the size of some countries are sacrificed for this.[181]

Taxes on mineral raw materials, suggested initially in 1983 for the entire European Union,[182] could be instituted too, relatively simply at the level of the importer and directly with the domestic producer. Their introduction would immediately generate a recycling impetus, and could help to reduce energy consumption significantly. This would also stimulate technological development to replace mineral resources with, for example, ceramic materials, glass fibre or vegetation-based raw materials.[183]

The next level of taxing environmentally damaging chemicals would also produce these effects: that is, a further recycling stimulus and more efforts to substitute environmentally friendly materials in a wide range of uses. In the packaging industry, it might bring about the use of foils made from plant matter that could be recycled back into nature. Chemical fertilizers would be replaced by crop-based fertilizers, made from water hyacinths, for example. Synthetic construction materials could be replaced by biologically based ones, and synthetic drugs could be replaced by those based on natural substances. Research and development in these areas would experience extraordinary stimulation, and waste collection and disposal problems would be reduced. Developing countries could switch from exporting nonrenewable minerals to exporting renewable crop-based raw materials. This form of taxation could be gradually expanded to an ever wider spectrum of other materials, beginning with those that are especially environmentally burdensome. The tax could be collected directly from the chemical manufacturer, making the chemical industry the cutting edge on the path to an environmentally oriented industrial society. The number of jobs would increase as a result of these substitution efforts.

In 1996, the German Federal Republic's total import bill for raw materials was about $22.3 billion, without including crude oil, natural gas, coal and uranium. Until now, customs duties have varied widely: for some categories, there are no duties at all, but for some minerals they are as high as 14%, and as much as 10% for non-ferrous metals. All imports are, however, subject to the general value-added tax. In contrast, corporate taxes totalled $48.3 billion in 1996.

Added to all this would be taxes on land use and water pollution, best addressed by local taxes. A necessary supplement would be a property sales tax to counter antisocial property speculation. Japan serves as a good example of the massive burdens that property speculation imposes on the entire economy and social structure. There, extremely high property prices are the most important factors of social entropy. Even in this sector, it would translate into practice the underlying principle of an entropy tax, namely that non-reproducible or finite goods are removed from the laws of the market. In the final analysis, the rules of the market cannot be used to deal with anything that is not reproducible and which therefore belongs to the collective good. Private ownership of property as such is not affected by this type of taxation, but business transactions involving property are, since they lead to disruptions of the sensitive social and environmental fabric, the destruction of residential lifestyles in cities, and to immense profits for a small segment of speculators who don't have to lift a finger to earn them.

For companies, the overall effect of an entropy tax would be a shift in focus of their efforts towards greater efficiency, which currently centres on the elimination of jobs, to the substitution of non-renewable resources by renewable resources and the avoidance of environmental burdens. To supplement such a tax strategy, decision-makers should ponder the idea of making imports of industrial goods from countries without comparable provisions more difficult by raising customs duties on these products. A large economic region such as the European Union could afford to do this. The strategy to save the globe is more important than the ideology of a free world market. Those national economies that insist that nature and mankind foot the bill for the real economic, social and environmental costs should no longer be permitted to enjoy market advantages as well.

This plan would thus correspond to the goal of stabilizing, redirecting and broadening the Industrial Revolution, while avoiding the damage it has done, and thereby humanizing it. The means are:

- raising the price of and saving more conventional energy while shifting to renewable energies;

- taxing those aspects of production that are finite, energy-intensive and environmentally damaging, but also offering in exchange relief on other aspects – that is, increasing the cost of energy and material consumption, but reducing other consumer costs;

- using new social gains while avoiding social cutbacks; redirecting industrial society's material flows from exploitation and destruction of nature towards opening up and conserving nature.

9

From a Nuclear Non-Proliferation Treaty to a Solar Proliferation Treaty

For decades, the societies of the southern hemisphere have been victims of the East–West conflict. Not only were large intellectual and material resources tied up in the East–West armaments race, but attempts were made to carve up the South into East–West zones of influence, and to impose one or other political system on that part of the globe. However, both Eastern and Western models have failed in the developing countries. That failure prompted frequent calls for a polycentristic, not bipolar, world order.

After the disintegration of the Eastern Bloc, the opportunity for self-determination in the Third World has become not larger but even smaller. Developing countries used to be able to play one world power off against the other and to let themselves be courted, but that is over now. Since these countries have drifted into ever deeper economic dependence, they are now subject to an exclusively one-sided blueprint for development. The only resistance has emerged in the shape of what is referred to as Islamic fundamentalism, mainly in oil-rich states that have, however, entered into a community of interests with the Western power centres. Since the decline of the Soviet Union these Western countries act as a monolithic power bloc vis-à-vis the South, just as they did earlier towards the East. Despite its lack of success, their development policies have become more uniform, although the entire situation cries out for a radical change of direction. At best they discuss the means of increasing development funds and easing exports for the developing countries, for example, but hardly ever the ends of a new development policy.

The Gulf War, which was supposed to secure the oil fields for the welfare of the West, had far more of an impact on the international politics of the 1990s than the global environmental crisis. The new 'extended security concept', originally a euphemism for the political reduction of social tensions, has instead led to extended plans for military intervention. Without any new goals, the politics of development will not be able to find a way out of the present dead-end situation, nor will it make any progress without new concepts for real action; instead the false starts that have already failed will be prolonged with ever more being spent on them. Only a solar development revolution can lead the way out of this labyrinth.

'Solar for Peace' instead of 'Atoms for Peace': the Solar Proliferation Treaty

The notion of deploying internationally an energy carrier outside the confines of the commercial energy markets is not new. It has been attempted for decades with the wrong subject, nuclear technology. Even today, the financial and institutional effort on behalf of nuclear power is much larger and more intensive than that for the international dissemination of solar energy. Organizations that represent these misguided efforts include the International Atomic Energy Agency (IAEA), which has the task of controlling the international nuclear fuel cycle and promoting technology transfer; the Nuclear Energy Agency, a subsidiary of the International Energy Agency (IEA) set up by the OECD member states (which nevertheless still do not see the need for a separate solar energy agency); and the EURATOM agency of the European Union. There is nothing comparable for solar energy, neither within the Western international community nor on a global scale. Proposals along these lines have, until very recently, been firmly rejected by international leaders.

Ever since the legendary 'Atoms for Peace' speech by President Eisenhower in 1953, there has existed the contradictory strategy of, on the one hand, preventing other countries from acquiring nuclear armaments while, on the other, supporting the proliferation of nuclear power plants even though it is clear that this support creates one of the preconditions for the ability to acquire nuclear weapons. The Nuclear Non-Proliferation Treaty took effect in 1970: basically, it stipulated that member states that do not have nuclear weapons abstain from their manufacture or acquisition. Those who have nuclear weapons, or have a head start in nuclear technology, promise – according to the treaty's text – to 'end the nuclear arms race in the near future and to achieve nuclear disarmament' and ' . . . to facilitate as much as possible the exchange of equipment, material and scientific and technological information for the peaceful use of nuclear energy'. The treaty is intended to contribute to the 'further development of the use of nuclear energy for peaceful purposes', taking 'into due account the needs of the developing areas of the world'. In terms of nuclear energy, the Nuclear Non-Proliferation Treaty represents both the greatest dangers as well as the greatest hopes of the 1950s and 1960s. The dangers have remained; the hopes have vanished.

The obligation to dismantle nuclear weapons has not been and is not being adhered to. The prospect of nuclear disarmament 'in the near future' was overtaken, after 1970, by an increasingly breathtaking nuclear arms race. Even after the demise of the Soviet Union, the Western nuclear states and NATO have rejected complete nuclear disarmament because they regard nuclear weapons as the ultimate deterrent, even aimed against states that have no nuclear weapons of their own. They refuse to acknowledge that this reasoning increases the dangers of international proliferation of nuclear weapons rather than reduces it, because it motivates some states to counter the threat of nuclear weapons that are targeted at them with nuclear weapons of their own.

The commitment to nuclear technology, on the other hand, was fulfilled because of the prospect of lucrative business deals. The results, however, lag behind expectations. One reason is the complexity of nuclear technology, which means that most countries cannot manufacture the 'product' themselves and therefore have to import it, something for which developing countries lack the necessary foreign exchange. On the other hand, for those who want to obtain an unofficial nuclear weapons-building capacity of their own via official civilian nuclear technology, the Nuclear Non-Proliferation Treaty is of some initial help. Thus there is the strong probability that, by invoking the Nuclear Non-Proliferation Treaty, exactly the opposite occurs of what had been intended. The 'Atoms for Peace' concept cannot be implemented because it is impossible in practice to separate totally civilian from military nuclear technology. Full, universal nuclear disarmament is therefore an imperative for world peace, and must include a prohibition on the dissemination of 'civilian' nuclear technologies.

The final article of the Nuclear Non-Proliferation Treaty states that a conference should be convened 25 years after its enactment to 'decide whether the treaty will remain in force indefinitely or whether it should be extended by one or several specific period or periods of time'. This period is coming to an end, and new ways of preventing a nuclear arms race by force are making their appearance, keeping the world in suspense and increasingly militarizing international politics instead of demilitarizing it. Such efforts to prevent a 'nuclear death' distract us from the efforts to prevent 'entropy death'. Only a new paradigm instead of the old nuclear one can liberate humanity: a paradigm that can be widely proliferated, that cannot be abused, and that can be made accessible to everybody. 'Solar for Peace' is the only appropriate guiding principle suitable for the use of energy on a global scale at all times. Mankind needs a Solar Proliferation Treaty. No measures will be needed against military abuse of solar energy because there is and can be no such thing as a solar weapon, either now or in the future; happily, the energy density of solar radiation is not sufficient to develop a weapon of mass destruction. In 1993 EUROSOLAR developed a corresponding draft proposal and presented it to UNESCO at a preparatory conference for a 'Solar World Summit'. It did not include a discussion of international politics, mainly because the professional specialists of foreign policy have too little to do with matters of the security of world energy. Such a treaty must include an international institution whose main responsibility is the transfer of solar technology.

The International Solar Energy Agency

The establishment of an International Solar Energy Agency was demanded in 1990 in a EUROSOLAR memorandum.[184] Many non-governmental organizations plus a few governments and political parties have supported this demand. In 1992 it became a central part of the recommendations of the United Nations Solar Energy Group on Environment and Development (UNSEGED), which had drafted recommendations on solar energy for the UN Secretary General before the Earth Summit in Rio. As early as the 1981 UN Conference on Renewable Energies in Nairobi, representatives from developing countries demanded the creation of a UN agency to support solar energy. This demand was rejected in 1981 by the industrial countries, and again in 1992, primarily by the American and Japanese governments, which shot down the suggestion for an International Solar Energy Agency in the preparatory conferences for Rio. There are a few UN development organizations that support solar energy programmes, but none of these has the authority that a Solar Energy Agency would have to have.

This agency would be mandated to assist any country to close gaps in its development of solar energy and build an independent infrastructure for its use, from research and training to production. This should include:

- assistance in setting up research institutes;

- construction of demonstration facilities in every country for the entire spectrum of new solar technologies;

- the exchange of scientific and technical information and training programmes for scientists, engineers, business executives and administration officials;

- consultation on establishing manufacturing facilities;

- dissemination of experience of launch programmes and planning such programmes;

- cooperation with governments, financial institutions, development and environmental organizations;

- execution of development projects.

The most important organizational task of such an agency would be to create technology transfer centres in numerous locations around the world, especially in the developing countries, with multiple tasks: development of solar technology applications; motivating, advising and training business executives and entrepreneurs, manual workers and local administrators; and carrying out regional projects.

It is not really surprising that the concept of solar technology transfer has been rejected so far. While the transfer of nuclear technology improves the market prospects of only a few power plant manufacturers, the opposite is true for solar energy technology: because it is less complex and free from danger, solar technology can be manufactured in developing countries themselves. Since transferring solar technology is thus not part of the global marketing strategy of industrialized countries, it is the precondition for an independent energy supply in developing countries. By rejecting the creation of such an agency, industrialized countries want to reserve for themselves a future market in whose expansion they have no real interest at present. If the industrial countries were to start manufacturing solar technologies on a large scale for developing country markets, then there would be no reason to delay the introduction of solar technologies to the home market. The energy-industrial complex believes essentially that, if the time is not yet ripe for using solar energy at home, it cannot yet be ripe for developing countries. If developing countries were to begin mass production of solar energy technology and could sell them to industrial countries as well as to their home markets, it would represent a serious challenge to the manufacturing monopoly of the rich industrial nations.

The need for solar energy, especially in the developing countries, is barely questioned any more. This explains why developing countries are regarded as future markets for this technology from industrial countries and why, therefore, the West does not want to nurture local competition. Its obstruction of the establishment of an International Solar Energy Agency reveals the unscrupulous egotism of the Western energy establishment, which is indifferent to the development disaster of the South, the approaching climate catastrophe, and the growing streams of refugees. It also reveals a dangerous short-sightedness: the developing countries cannot be big markets for solar technologies because they do not have the purchasing power. Either they will produce most of their solar technology themselves, or Western sales will amount to the equivalent of a few drops of water in the desert sands.

The successful resistance so far to such an agency also reveals how tightly the political, economic and scientific elites of the developing countries are interlaced with those of the industrial countries. It is difficult to explain otherwise why time and again developing countries let themselves be persuaded to stay away from solar energy, or, rather, why they don't opt for solar on their own.

Obviously, the reasons cited here for blocking a Solar Energy Agency are never officially confirmed. Instead, so the argument goes, it would make little sense to create yet another international agency; existing ones could or should handle this additional task. Yet none of these existing organizations has been given either the required mandate or the funding. Another poor excuse, contradicting the first one, is that the poor performance of existing UN organizations discourages the formation of a new one. Surely, if their work is so unsatisfactory, that should be an even better reason to create a new, efficient institution.

The time is overdue for the majority of states not to let themselves be prevented from recognizing and acting on their own interests by a minority of oil-producing and industrial countries. It is absurd to make the establishment of an International Solar Energy Agency dependent on the agreement of those governments who do not want it in the first place, and whose involvement would only do damage because they would obstruct the agency's efforts. Such a Solar Energy Agency could be the link for a cooperative alliance between the developing countries and the enlightened industrial countries. Whenever international economic associations have been formed in the past, their initiators did not ask whether there would be objections from those whose economic interests might suffer. For all these reasons, some governments should go ahead and launch such an agency on their own, without waiting for a general consensus that will never come; the Austrian Chancellor Vranitzky hinted at such a possibility in 1992.[185] The agency would have to remain open for other countries to join subsequently.

Without an international agency, the necessary transfer of solar technology will remain dependent on more or less fortuitous bilateral cooperation agreements among individual states. With such an agency, the economic blossoming of solar technology would no longer depend on competition and jealousy among global economic competitors – who have pretty much outdone each other in stifling solar energy. With their minimal activities, these go-slow specialists with small energy departments keep one foot in the door of a future solar market, but with the other foot simultaneously attempt to slow down market development. If the solar market takes off without them, they want to make sure they are part of the action. However, solar energy technology is too important for mankind to let it be subject to that sort of gamesmanship any more. With the help of technology transfer via the Solar Energy Agency and widely distributed solar technology centres, every effort must be made to ensure that all countries become solar technology producers of their own in the shortest time possible. Only then will the developing countries be in a position to satisfy their solar energy needs. Once they produce solar technology of their own, they will have the chance to become exporters of industrial products, thus advancing beyond the role of mere exporters of raw materials and agricultural products, a process that chips away at their natural assets as they are forced to sell foodstuffs they urgently need to feed their own people.

Precisely because the creation of many of the initial facilities for an energy supply system in the world's southern regions are at stake, the prospects for launching solar energy are greater there than in industrial countries. The task of a Solar Energy Agency would also include creating a system of patent infor-mation covering solar technology and offering patent counselling in this field; helping to arrange licensing and joint-venture production; and drafting uniform industrial standards for solar technology to facilitate global interaction (see Figure 6 on page 222). Solar technology transfer is an alternative to development credits and services that so far have focused mainly on the use of fossil energies. Independent and unhindered economic development of the South is impossible without solar energy.

From the 'Green Revolution' to a Solar Revolution in Developing Countries

The Solar Energy Agency would also be the appropriate institution to help shift the focus of international development aid efforts towards solar energy, and to coordinate its organization. Part of a successful switch should be to give priority to those applications, within the broad scope of solar energy use, that are most important for the survival of rural populations in developing countries:

- solar-powered water pumps and irrigation systems, to exploit fully the agricultural potential (in India, for instance, frequently only one harvest is possible because of water shortages or an inability to store water in places where two and even three harvests are possible);
- operation of agricultural machinery with locally manufactured fuels, exploiting the availability of natural

fertilizers, and using the ability to dry and store harvests with the help of solar energy;

- cooking and electric lights;
- availability of energy for small-scale manufacturing.

Table 20 shows which solar technologies could be used to meet various social needs.

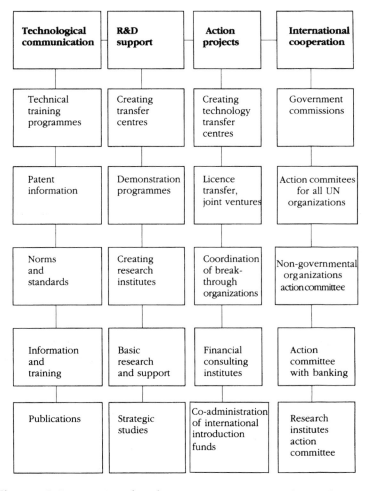

Figure 6 *International Solar Energy Agency: tasks and management structure*

Table 20 *Solar technology applications in developing countries*

	PV cells	Solar high temp.	Solar low temp.	Small hydro	Wind	Bio-mass combust.	Bio-mass gas-ific.	Bio-gas	Bio-fuels
Cooking			x			x		x	
Drying, water and space heating			x			x		x	
Telecommunications	x			x	x				
Domestic electricity	x			x	x		x		
Sterilization and disinfection	x		x					x	
Desalinization and distillation	x	x	x			x		x	
Water pumps	x	x	x	x	x	x	x		
Compaction, threshing and forest cutting	x			x	x		x	x	
Cooling	x	x	x	x	x		x	x	
Tractors				x	x		x	x	x
Process steam		x	x			x		x	
Cogeneration	x	x			x	x	x	x	
Large-volume harvesting		x				x		x	

These solar energy opportunities must be introduced rapidly and widely because only they can slow down population growth and halt refugee movements. A change both of methods and priorities in the politics of development aid is required to achieve these aims and to halt global climate catastrophe.

Crash programmes are necessary in the politics and policies of development: carefully targeted global projects that can be carried out in short order. A large proportion of the world's population exist without electricity in developing countries. They cook and heat with wood or dung, depriving themselves of valuable fertilizers for agriculture and wastefully exploiting their natural vegetation. They have to travel further and further to obtain firewood. According to World Bank reports, about 1 billion m^3 of wood is burned every year for cooking alone. Together with the dung burnt and crop waste, this corresponds to an amount of energy equal to 5 million barrels of oil per day and an average of 500 km^2 (193 square miles) stripped every day of wood or 180,000 km^2 (69,500 square miles) per year, almost 0.5% of the world's wooded surface. The forests and trees don't always disappear completely, but even if half of that recovers and grows again, the annual acreage lost would still be roughly the same size as Austria.

The firewood crisis in developing countries is assuming dramatic proportions. According to studies by FAO (the UN's Food and Agricultural Organization), there will be more than 2 billion people without enough firewood. The flight to the cities, caused largely by a shortage of energy in rural areas, in turn aggravates the energy crisis because easily transported charcoal is used in the cities as well, and the overall energy losses are higher than if the wood had been burnt directly in rural areas. Every person migrating to a city thus doubles his or her energy consumption – a truly vicious circle.[186]

The simple introduction of solar cookers for 500 million families would be sufficient to solve the firewood crisis. Manufacture of an appliance that could cook 2 litres of food, keeping it warm for several hours, would cost about $62.50, including $40.60 in materials costs, when produced in low volumes of a few hundred. Manufacturing these solar cookers in developing countries might halve the cost. Assuming the costs to be $31.25, including manufacturer's profit and marketing costs, equipping 500 million families with the device would cost $15.6

billion – about 2.5 times Germany's annual budget for development aid, or two thirds of the total cost of developing the European fighter jet, or a bit more than the German government's contribution of $11.25 billion to the Gulf War.

For this expenditure, solar cookers could save the firewood otherwise needed by 2.5 billion human beings, would counter land degradation and combat the greenhouse effect, and could slow down mass migrations and the flight into the slums of cities in developing countries, which already have millions of inhabitants. It would probably even slow population growth itself, because the more human energy is needed to obtain wood, the more children are produced.

If these cookers were to be produced in developing countries several million jobs would be created in the trading operations and small industries. It would be neither necessary nor sensible to give these cookers free to all families, making the entire project even less expensive. However, it would make sense to distribute them free to the poorest of the poor, who have no purchasing power at all. There should also be a comprehensive plan to create craft production centres in villages as a sensible introduction strategy. Crash programmes of this type would need cost controls, achieved most easily by setting up competing production centres. On the other hand, programmes of this type are practicable only if the widespread, patronizingly pedagogical, development aid mentality is overcome, according to which – exaggerating a bit – every solar cooker needs continuous care, control and sociological analysis.

A much more far-reaching crash programme would involve the introduction of photovoltaics (PV) in developing countries, a strategy that would lend itself to building a new economic North–South relationship. At present, there are a few thousand solar-powered water pumps in developing countries, as well as a few village power plants, inoculation stations and power plants for schools or rural telephone networks. Photovoltaics would enable the generation of a minimum of electricity to improve comfort levels in rural areas of developing countries without requiring power grids. The idea of a Global Photovoltaic Action Plan was originally suggested by Helmut Glubrecht, founder of Lower Saxony's Solar Research Institute, and was developed in more detail by Wolfgang Palz of the EU Commission. From this idea arose an action plan called 'PV for the World's Villages' in

1996. It was developed for the EU Commission by the Agency of the European Renewable Energy Research Centres (EUREC Agency) in collaboration with EUROSOLAR, the engineering consultancy WIP, and eight institutes and firms. It investigated the measures required to provide electricity to 3 billion village inhabitants. As a first step, it aimed at an average capacity per person of 10 Wp using photovoltaics. This would result in an average daily amount of electricity per person of 50 kWh. The total costs of the programme would be about 150 million ecus spread over 30 years. It would rapidly boost the mass production of photovoltaics, above all with facilities located in developing countries. It would create a new installation and maintenance service sector, and technicians would gain and disseminate experience.

Such a widespread, crash programme and framework for global activity would represent a solar development revolution, modelled on the 'green revolution' of the 1950s. That programme focused on increasing the world's food production to fight global hunger. It achieved a number of successes, but most were short-lived and had a number of negative side-effects, which qualified the successes. Wheat, corn and rice yields were increased, but at the price of a massive increase in oil consumption in developing countries; and of loss of biological diversity and the unwise use of nitrogen, phosphorus, potash and insecticides. This approach copied agricultural practices in the industrial countries, characterized by an idiotic energy balance. The energy gain in the form of foodstuffs is offset by a higher use of energy, leading to a collective march into a dead end.

Further, energy costs have a more negative economic effect in developing countries than in industrial countries, as they have accelerated the displacement of small farms by large-scale operations, backed by First World capital, and the shift to food exports has frequently worsened the food supply for the local population. Agricultural 'modernization' has cost countless individuals their jobs.[187] With a solar revolution, this trend can be reversed by creating an autonomous energy supply system independent of primary energy markets. The combination of renewing vegetation and agricultural initiatives, described by Peter de Groot, Alison Field-Juma and David O. Hall using the example of Kenya's semi-arid areas as 'Taking Root',[188] represents a huge opportunity for the utilization of solar technology for irrigation and the cultivation of food, energy and raw material crops. In forest regions, on the other hand, the emphasis might be on a mixed forest/agriculture economy in which the forest is used as a source of energy, for food products and for commercial and industrial purposes. A solar energy supply system is thus the basis for assuring existence and small-farm employment in Third Word countries. Also, it would not only halt the flight from the land, but also stimulate an organized flight from the cities, a voluntary departure from the slums, once a return to rural areas promised better living conditions.

Re-Greening the Earth

It is possible over the next half century to bind by reafforestation those amounts of CO_2 that nature will no longer be able to absorb. It is therefore no longer acceptable to aim for only a gradual reduction of climate-damaging trace gases, or even mere stabilization of these emissions at the 1990 level. It is a syndrome of our time that those who regard the continuation of the current environmental war at present levels, rather than its further escalation, as an environmental protection programme are treated as progressive, deserving and expecting public applause.

There is no rational reason not to start immediately on a large-scale re-greening of the planet, although there are doubters who demand environmental impact assessments before starting such a programme, while simultaneously the combustion of fossil energies briskly continues. A programme for re-greening the earth would give mankind sufficient time to put into action the conversion of the energy systems, which also cannot be delayed. Global reafforestation can get under way immediately because it does not need any technical developments or production facilities: the soil is the production facility. To make re-greening as cost-effective as possible, it should be put in place in developing countries and in the countries of the former Soviet Union. Large areas are available for that purpose: Russia with 17 million km² (6.6 million square miles), China with 9.6 (3.7), Brazil with 8.5 (3.3), India with 3.3 (1.3), Argentina with 2.8 (1.1), the Sudan with 2.5 (1.0), and Mexico with 2 (0.8), to mention only the largest. This would create jobs for millions of people, it would upgrade nature and land in these countries, and it would create a vast forest industry that, as an economic factor, would contribute both to the commercial energy supply and to the harvesting of natural raw materials, which in turn would ease the transition of the global economy to one based on biological materials.

Such a global reafforestation programme could be based on international re-greening agreements. Developing countries would contribute land and manpower, and the industrial countries would add the necessary capital. Industrial countries should pay for reafforestation as they are the principal cause of climate change, with their excessive energy consumption. They would thus start to repay the environmental debts to humanity that they have incurred in the false hope that they would not have to settle up. Such a financing arrangement would be tied to a specific purpose, and its effect would be quantifiable and controllable.

This programme would not need an international agency to implement it because reafforestation could be handled through direct, bilateral agreements with individual developing countries. The financing countries would merely have to show that they had paid their appropriate share to the reafforestation effort within a prescribed period of time. By agreeing to pay for reafforestation of an area that corresponds to its share of energy consumption, an industrialized country could monitor the operation's cost-efficiency. The overall goal should be the regrowth of 10 million km^2 (3.9 million square miles) of forests in a 10-year programme, which would absorb at least 10 billion tons of CO_2 annually over at least 40 years.[189] One condition should be that these agreements are signed only with developing countries that simultaneously halt destruction of their forest reserves. This would provide a genuine incentive to preserve tropical rainforests.

Assuming that the entire programme costs a maximum of $313 billion, the shares of the total effort for each industrial country can then be clearly defined. If each country's share is based on the ratio of its gross domestic product to OECD's total gross product, we arrive at about 8% for Germany, or about $25 billion: about $2.5 billion per year. For Great Britain and Italy the respective shares would be about 6% each or $1.9 billion; about 7% for France or $2.18 billion; about 20%

for Japan or $5 billion; and 30% for the United States or $7.5 billion. These costs represent upper limits. 'Global Releaf' has calculated that in some cases the cost could be reduced to as low as $30/ha (2.47 acres), especially if 'sweat equity' could be mobilized in developing countries, such as voluntary human labour or the use of armed forces. Each country would be free to find the best way to raise funds: by reducing aid contributions from the national budget, by – intelligently – buying forest products from the reafforestation areas, or by providing aid workers for forest cultivation or sending in 'green troops': that is, involving only small or no transfer payments. After all, nothing less than preserving the climate is at stake, with further benefits such as social reform in developing countries as a bonus. An initiative such as this will probably cost members of the European Union less than they will have to spend in the foreseeable future to house refugees from Africa, or on 'rapid deployment forces'.

The largest single reafforestation project at present has been developed and costed by the Chinese government. This project envisages the reafforestation of 74,000 ha (183,000 acres) in the upper and middle reaches of the Yangtze River. This would require international aid of about $100/ha (2.47 acres) in addition to the Chinese government's own efforts, contributions by regional administrations, and voluntary help from farmers. The expected result is the absorption of, on average, more than 15 tons of CO_2 per hectare per year. The global effect would be a reduction of more than 100 million tons of CO_2 per year for a price that would not even pay for two modern jet fighters in the West. One bonus, among others, would be the reduction of soil losses caused by the erosion of an estimated 400 million tons annually of the region's soil.

The suggestions for re-greening the earth sketched here, and the numbers quoted, offer clear advantages compared with the many other suggestions so far put forward, none of which has yet amounted to anything. This re-greening initiative can be launched without the agreement of all governments on a common goal beforehand! Each can assume its own share of the overall responsibility for averting a climate catastrophe, and can implement reafforestation with one or several partners. The old excuse that one would be willing oneself, but that unfortunately 'the others' don't agree, won't work any more! When a government spends large amounts of money on other programmes that it deems essential, it usually does so independently of the actions of others. If one or several governments went ahead with such a programme, it would put the pressure of world opinion on the rest.

The alibis of the past have lost their credibility in the face of present needs and the feasibility of re-greening. The willingness to take at least these measures is the litmus test of environmental credibility for every government. Those who are unwilling to contribute to the easiest, fastest-working and most cost-effective initiative are not going to participate willingly in the more difficult, slower and more expensive measures: they will, morally, have lost all legitimate claim to political responsibility in our time.

The Global Action Fund

In addition to national initiatives, global action is required. This means funding. There is a concrete example of how more than $70 billion was raised around the world in a matter of weeks for a global action: financing the Gulf War – officially to rescue one nasty government from invasion by another nasty government. Despite long and laborious haggling at the Rio Conference, it proved impossible to scrape together anything near an equivalent sum of money for additional measures to save nature and the poorest of the world.

A number of models have been suggested in the past for the creation of an international fund to finance measures for protecting the planet's atmosphere in those regions of the world unable to do so on their own. The key element is an international CO_2 tax based on fossil fuel consumption, partially augmented by an equivalent levy on nuclear energy and imposed either by industrial countries alone or by all governments. All these proposals have been rejected. Governments have refused to give equal treatment to issues that in their view were not equivalent, even assuming they considered such ideas as worthy of discussion in the first place. With some justification, they regard as unequal the various positions of the industrial countries themselves. This disparity, among other factors, has so far caused the failure of the proposal for an energy tax within the

European Union, although the revenues from such a tax were not even intended for an international fund.

The most important reason for rejecting such a tax based on CO_2 emissions – or any other criteria – is obviously that nobody wants to give up anything. The financial difficulties of most countries have become too serious. The fear that it might create economic disadvantage and favour those who refuse to contribute looms large. Great, too, is the scepticism towards a 'super-fund' that others might be able to tap into easily, and which would create a new, impenetrable bureaucracy. A few arguments are informed with a pragmatic logic, but they begin to wear thin if there is no creative vision of an alternative concept to allay such misgivings.

The basic idea of an international fund for global action is absolutely clear. The present situation, in which international organizations have barely enough money to keep themselves afloat, degrades them to mere ends in themselves. Funding sources have to be found, therefore, that provide sufficient means for global action and that make possible a fair and equitable distribution of the burden. A general CO_2 tax, or a tax imposed only in industrial countries, unfortunately has no chance of success at the moment: if governments levy a general additional energy tax, experience shows they will spend the revenue on themselves.

These observations lead to the concept of a global action fund that would obtain income from a sector that has not yet been taxed, and which is directly linked to climate dangers: civil aviation. Strangely enough, international air traffic has played only an incidental role in the climate debate, although aircraft emissions cause serious climate damage – and the trend is growing. As soon as an aircraft reaches the stratosphere – an altitude of more than 8,000 m (26,250 ft) – the emissions have a much more damaging effect than at lower levels in the atmosphere. Furthermore, air traffic is increasing rapidly, especially cargo and holiday flights. Emissions from air traffic are approaching 10% of all fossil energy waste gases. In 1997 there were 600 million flight reservations quoted for the year on the International Tourism Exchange in Berlin, and they are expected to rise to 1.8 billion over the next 25 years. Flying is much too cheap, particularly as it is internationally tax-free! The burden borne by every car driver does not bother the manager flying from Frankfurt to New York or Tokyo to Paris, nor the tourist flying to the Maldives or the Caribbean. Transporting milk by road carries the burden of fuel tax, while the transport of orchids by air from Colombia to Europe does not. Even if there were no environmental problem, this most blatant of all favouritism scandals should be stopped immediately. There was public uproar in Germany in the mid-1980s when the government attempted to abolish the aviation fuel tax for private pilots, yet the fact that airlines already enjoy this tax-free status should have been worthy of a much bigger outcry.

Nor is the general tax exemption for international airlines an old regulation that has not been abolished simply because of an oversight. It was in fact enacted at a time when global climate dangers were already known, when it was evident that forests were dying because of acid rain, and when all countries were discussing increased fuel taxes for motorists: in the autumn of 1986 at the 26th general assembly of the International Civil Aviation Organization (ICAO) in Montreal. The German government answered my parliamentary question about the legality of this decision as follows:

> ICAO decisions are not immediately legally binding. But Articles 37 and 38 of the ICAO agreement urge member states to follow its standards. Deviations within a country have to be reported to the ICAO. The decision to exempt (airlines) from taxes on mineral (crude) oil is being generally adhered to by ICAO member states.

Part of the revenues derived from a tax on aviation fuel should be set aside for a global fund for environment and development. It would not detract from current government budgets, and would tax a highly internationalized sector that, for reasons of tax equality and global environmental protection, should be taxed as a matter of principle. Further, this tax would not impose burdens on socially underprivileged population strata. If the transport of orchids becomes more expensive or disappears completely, humanity won't suffer, and making some holiday flights more expensive would not be unreasonable either, as the existing taxation burdens on the general public are already much more intolerable. The argument that air travel will revert to being a privilege for the rich does not wash; it is of as little social relevance as arguing that everybody should be entitled to a cheap Porsche.

Fuel taxes on high-octane petrol are very different in different countries. The average fuel tax per litre in the OECD countries is about $0.50. Placing a tax of $0.50 on each litre of aviation fuel would produce annual revenues of more than $100 billion, based on current consumption. Even assuming that every OECD member country taxed aviation fuel at the same level at which it taxes high-octane petrol for cars, this would still produce annual revenues of more than $70 billion. If we assume further that some countries, the United States, for instance, did not initially go along with the idea, others – such as the European Union – could still launch the new tax programme because each aeroplane has to be refuelled where it takes off. In other words, there would be no danger of damaging competition.

In any event, it should be part of a global environmental policy to respond to those countries that refuse to take part in necessary international measures with trade sanctions in other areas. This would be possible if the initiators of such a global environmental policy were economically strong and powerful enough, as the European Union would be if it applied its economic influence in the right places in the right way. Half of such fund monies could then be used for projects of the type described in the preceding sections. Even if this sum turned out to be lower, owing to the likely fall in airfares, than it would if airfares remained constant, there would still be enough money to fund the global initiatives.

A Pioneer Corps and Rapid-Response Forces for Development

Without a carefully tailored transformation of economic structures, there is just as little chance of real change as there would be without modifying the functions of political institutions. Once the intrinsic usefulness of these institutions is in doubt, or they have become superfluous, they should be dissolved or given new tasks. It is absurd that NATO states are frantically scrambling for new military tasks for their armed forces since the end of the East–West conflict, and are fixated on keeping busy with new intervention doctrines and troops. Many developing countries are accused of wasting scarce funds on their armies instead of using them for economic development, yet the West is doing exactly the same.

In developing countries, armies are frequently the only government institution capable of genuine action because they are trained, organized and equipped to an above average degree. Most of the time they serve more to guarantee the domestic stability of their regimes than to assure external security. Since they have few other duties, too often they regard it as legitimate to assume power in the country at times of crisis. It would be naive and idealistic to assume that developing countries can be persuaded to reduce their armies, especially if advice of this sort comes from the heavily armed industrial countries. It would be more realistic to employ constructively the potential represented by these armies for economic and social construction – in other words, for the peaceful resolution of the underlying causes of social unrest rather than suppressing that unrest by force.

The idea of using military organization and technology potential for constructive purposes is not new. The Roman Empire used its legions to cover Europe, North Africa and Asia Minor with an infrastructure that included streets, bridges, dams and dykes. Another remarkable example in more recent European history was the engineering corps of the Napoleonic army, which was employed to build public infrastructure and which in a short period of time achieved noteworthy feats of construction impossible with civilian forces at that time.

The underdevelopment crisis in developing countries has become so widespread and accelerated to such an extent that it is impossible to continue the previous policy of simply increasing funding for development. The potential for social productivity of otherwise unproductive and expensive military forces must be put to use in the campaign against environmental and development catastrophes, and they must be mobilized for civil works. This is also true for the deployment of armed forces from developed countries to aid these regions, not only after conflicts and catastrophes flare up, but also to prevent them. It applies in much larger measure to the armies of the developing countries themselves. They should be organized in such a way that civil works are no longer an exception, but part of their regular mission. The tasks at hand are things such as reforesting rather than defoliation, the removal of water pollution rather than placing mines, constructing bridges and roads rather than destroying them, building dykes and irrigation systems instead of causing floods, fighting forest fires rather than napalm bombing, and decontaminating land rather than poisoning it.

Developing countries lack the means to achieve a widespread and rapid build-up of economic infrastructure, especially in rural areas. There is usually a lack of money, personnel and the technical capabilities needed to create such an infrastructure. But the only halfway-efficient government institution with anything like the requisite technical equipment, administrative capabilities and moderately trained technical personnel is hardly ever used for these purposes. An army has the capability to build roads and bridges, and to clear land and waterways. Soldiers can dig trenches, and therefore could build irrigation systems. They have experience in solving transport problems under the most severe conditions. Their mechanics are trained in the maintenance of equipment and vehicles. Above all, they have a great deal of experience in organization and logistics, which can make a vital contribution to building facilities in remote regions.

Special assistance units from developed countries must be deployed to aid these efforts: rapid-response teams to help construction in rural areas. Their training and organization has to be designed to meet these needs. Every major formation of the armed forces should include a unit on call, equipped with a basic personnel team and equipment for humanitarian and development assignments. In deployment and training missions, this unit would have to be augmented by other personnel, depending on the type and magnitude of the mission.

The core of such troop units should consist of well-trained medics, supply and pioneer units, supplemented by marine protection units from the naval forces and efficient transport units. All would require appropriate equipment: ambulances and containerized medical facilities, mobile catering and sanitary facilities, equipment for road, dam, trench and bridge construction, for building fire breaks to protect against forest fires, for the decontamination and detoxification of soil and stretches of water, and for surveying tasks. The basic purpose of using armies is not to expand them, but to use the existing potential.

This will not do away with the need to create civilian organizations as well, for pinpoint deployment in paramilitary fashion. The idea of the late United States President Kennedy to establish a Peace Corps provided a model for many countries to organize development aid service corps of their own, staffed by volunteer aid workers. However, the worldwide mood then of a fresh beginning died with the atmosphere of pessimism created by the Vietnam War, and its early promise never returned. Today, the need for organizations of this type is greater than ever before, and there are many people willing to fight poverty, hunger, disease and catastrophes in unfamiliar climatic conditions. Among other things, the Peace Corps has made it clear that effective help does not necessarily require academically trained development experts. A high degree of motivation and organization plus a modicum of technical know-how are sufficient for the most urgent assignments.

The model for the American Peace Corps was, in turn, the Civilian Conservation Corps, created as part of the employment strategy of President Roosevelt under the leadership of George C. Marshall. Its tasks lay in the national arena – reafforestation measures as well as measures in soil conservation and the construction of fire breaks. Between 1933 and 1942, more than 2 million young Americans served on average nine months in this corps. American flood control legislation requires the US Army Corps of Engineers to render assistance in the event of flood catastrophes. It is trained, has the authority, and is equipped for the task. The Corps' entire structure is oriented to these civil tasks. In France, the country's 8,000-strong military fire brigade is used to fight forest fires.

It is an echo of the bureaucratization of political institutions that operate within tightly compartmentalized areas of responsibility and with abstract goals if points of departure such as these are not turned into reality and expanded.

10

The Energy of the People

'All human beings are born free and equal in dignity and rights. They are endowed with reason and conscience . . .' These initial words of the General Declaration of Human Rights formulate the most fundamental humanitarian obligations, yet the present dominating energy system does not, and cannot, satisfy these obligations. That the connection between the energy system and the attainment of human rights is unclear to many is due not only to the current situation, one in which only the word 'free' registers in the context of human rights but not the terms 'all human beings' and 'equal', nor the social foundations of freedom. In an insane way, the energy problem is simplified to a question of overhead costs. With reason and conscience cast aside, the antisocial and future-destroying consequences of such a view are ignored. The freedom of economic and cultural development, enjoyed by a minority of mankind, causes pauperization of the majority – and will, in the future, cause the pauperization of all mankind.

Since individual human beings cannot pick and choose among energy systems, a car or aeroplane journey creates a guilty conscience. The younger generation, in particular, is now denied a life free of anxiety because of the destructive consequences of the energy system, and the resulting mental and emotional costs have become incalculable. Liberation from this fear is conceivable only in one of two ways: either by accepting the system's consequences and thereby the increasing ruthlessness and brutalization of human relations – as if those privileged by today's energy system had the right to be the sole and the last surviving human beings on this earth – or by consciously and decisively reorienting the energetic foundations of society to the sun's powers. Precisely because energy is the most fundamental of all issues, the existing energy system violates the very basis of human rights – and the longer this continues, the worse it becomes.

There is an obligation to attempt everything possible not simply to mitigate the worsening energy catastrophe, but to prevent it. Nevertheless, the importance of solar energies to society would be fundamental for human development even without the global environmental crisis and without the limitations of conventional energy resources. Clearly, the utilization of solar energy is economically and socially the fundamental innovation for the global community. Its rapid introduction consequently represents a unique opportunity to:

- integrate the economy and thus redirect evolution's deadly detour, exemplified by the fact that this essential union has been turned into its antithesis, with fossil and nuclear energy as the most significant and most dangerous examples;

- slow down population growth and prevent international migratory movements;

- permit economic development in developing countries with the widespread use of nature's solar energy supply, and not simply a bad imitation of the North's own industrial growth;

- create new production structures and markets in which all economies and all people can participate, and which can be driving forces for the creation of innumerable new industrial and skilled production jobs;

- reverse the, until now unstoppable, trend towards ever-increasing centralization of economic and political structures and the attendant bureaucratization and social inflexibility;

- reduce international inequalities and make energy autonomy possible for every nation. Based on the direct connection between disposal of conventional energy, capital accumulation, extremely uneven distribution of wealth within the world economy and international dependency, a new interrelation between solar energy, widely distributed capital, a more equitable distribution of the world's resources and economic partnership might arise;

- prevent continuous and increasing tensions, crises and conflicts, which in many cases are caused by the unequal availability of conventional energy sources;

- prevent further global energy crises of the type experienced in the 1970s. Another energy crisis would finally plunge the greater part of the developing countries into a horror-filled delirium. For that reason they have no other choice but to take immediately the direct road away from the age of oil and towards the age of solar energy;

- overcome traditional social hierarchies as well as discrimination against women in developing countries, thus providing a viable sociological basis for democratic development based on human rights. Both of these are linked to the de facto energy-slave status, which in turn is caused by a shortage of energy;
- offer mankind a prospect of survival that would overcome its spreading fatalism about the future, and which in terms of social psychology would create a new mood of societal motivation.

To reject these opportunities merely because of the currently extra – in some cases only minimally higher – costs of a few cents per kilowatt-hour or per gallon of fuel is nothing less than sheer lunacy on the part of monomaniacal political and economic leadership elites. They represent an 'economy of death', as the American arms critic Richard Barnet calls it. Neither their increased power nor their elitist arrogance can distract attention from that. Solar energies, on the other hand, promise an 'economy of survival' – environmentally, economically and socially, in terms of developmental and industrial policies and politics. With them, it will no longer be a matter of 'faster, higher, further', which has both fascinated and demanded too much of mankind in this century, and not only in competitive sports. Instead, it promises to make life as we live it more compatible both with the natural world and with human nature, linked to a generally higher standard of living and, all in all, an improved quality of life. Solar energy is the energy of the people.

Exploding Contradictions

In the twentieth century, mankind had the best opportunity ever to shape society in a human way, but for the most part the chances were used badly or short-sightedly. All the positive achievements have been short term rather than long term. Democratic ideas have triumphed, but increasing social problems are now leading to their decline. Colonialism was overcome, but international exploitation has increased. Unimagined advances in efficiency and rationalization have been achieved, but lavish wastefulness occurs precisely in the strongholds established by these advances. Incredible discoveries have been made in technology and the natural sciences, but the dangers created by technology and the destruction of nature are more dramatic than ever before. For the first time in history, the North American-Japanese-Western European trilateral constellation embodies an economic, military and cultural concentration of power that encroaches on the entire community of nations; but nonetheless it is incapable of giving directions that could prevent a future crash. None of these contradictions can be blamed on scapegoats. They exist because the paradigm – the orientation model that determines scientific, economic and political thought and action – has a fundamental cardinal flaw.

The free market economy as guiding image of the Western political systems has unmistakable advantages in terms of innovativeness, competitiveness and consumption and, thus, of social functions as well for those who have the opportunity to participate in the market. But in principle it is blind to the external social and environmental consequences of its processes. These consequences cast increasingly long shadows, the more successful a market economy becomes. In and of itself it is incapable of distinguishing between reproducible and irreproducible, between socially useful and socially damaging values. While these values are indiscriminately commercialized, a social explosion will occur at some point. This is the basic conclusion derived from the Entropy Law, which as has been shown applies to economic, social and administrative processes as well. Fossil and nuclear energy, mineral raw materials, tropical forests, soil, air and water are not replaceable.

It becomes even worse if the consumption of a non-renewable good also causes social damage for other non-renewables. Since, especially under the current energy system, the non-reproducibility of resources coincides with social damage inflicted on other non-reproducible basic human goods – soils, air, water, the earth's atmosphere – the core of this basic flaw lies here. The free market economy is creating a disaster for mankind via the existing energy system, because it treats irreplaceable human goods as commodities that it can squander.

The unique advantage of solar energy is that it can take over the indispensable economic function of traditional energies – in economic structures transformed by solar energy, to be sure – and make it accessible to all mankind. Of the various non-renewable and environmentally damaging elements of the entire production process – conventional energy, mineral raw materials, chemical commodities – it is possible to replace entirely at least the energy component with a complete, inexhaustible and environmentally sound substitute. This is far from true for the other non-renewable elements, and more intensive efforts will be needed to replace them to any large extent. This will be the great research and development task for both public institutions and private enterprises in the coming decades. This then is the bottom line: today's opportunity to replace non-renewable and socially damaging energy carriers with solar energies offers the first and best chance for a new course in development. Inescapably, the utilization of materials and chemicals will increasingly have to be restricted. Assuming that there is a shift to solar energy, they will no longer be necessary for the energy supply.

Only a solar energy system is compatible with the positive aspects of a free market economy. As long as the full significance of that goes unrecognized, and as long as it is not acted upon, mankind will go after these problems in full cry, and will make innumerable efforts to solve them, but despite all efforts will see that they continue to grow – and still will not understand where the real, deeper problems lie.

The pre-eminent significance of these global challenges has been recognized, but new nationalist and fundamentalist movements are rampant. Much is said about the social functions of business and the economy, but the elimination of social responsibilities under the banner of improved efficiency and productivity increasingly determines economic decisions. Decentralization is recognized as a necessity, but the real control is becoming more and more centralized. Innovation is part of every programme, but the old priorities remain fixed. Lip-service is paid to the need to think in terms of interdependence, but action still takes place, unchanged, in small compartments. Long-term thinking is commended, but the reality betrays an increase in short-sightedness. Society regrets the lack of new fundamental political initiatives, and grows weary of politics. An uneasy degeneration begins to manifest itself – the most fatal of all dangers, because it constantly diminishes the likelihood of an alternative.

The calls for alternatives mount, but they have now lost credibility, because the mechanisms of the victorious Western mega-machines are believed to be, in principle, immutable. Thus, despite devastatingly misguided developments, the chorus of complaints falls on deaf ears, and truly alternative programmes are treated with ironic condescension or pushed carelessly aside. The alternative forces are battering against the padded cell walls of a Western high culture transformed into a lunatic asylum, a culture that doesn't acknowledge its gradual fall because others have fallen faster.

The strategies offered at present by politics and economics betray a paralysing fear of deliberately bringing about the necessary changes in the established order to escape the real risk of unwanted breakdowns with uncontrollable results later on. But the dominant decision-making patterns and decision-makers will be superseded eventually, one way or another. The key question is solely whether events will then move in the right direction, or whether the approaching chaos will accelerate, leading ultimately to an irreversible lack of alternatives. Reactions to the crises of the 1980s and the early 1990s show that we are currently in a dead end and unable to turn.

The Supporters of a Solar Strategy

It is impossible, however, to construct an alternative out of the very components of the development that is to be replaced. The solar revolution needs new supporters in society, in politics, in economics, in science and in the media. They need to combine to aim at an urgently required, motivating, realizable and promising perspective, and they must take the offensive in their pursuit of this goal. This perspective must be consistent – free of contradictions in its strategic concept – and it must concentrate on the essential and maintain freedom of choice. If the commitment becomes lost in minor details, the entire effort will turn into a Sisyphean labour, and will eventually be abandoned.

So far, discussion about solar energy has been conducted, even by many of its supporters, far too defensively and faint-heartedly. In tones that occasionally verge on the apologetic, supporters press for its acceptance as if they have to justify themselves – by rights, the position of those who refuse to introduce solar energy. All too frequently, solar energy is promoted as an alternative for the provision of small, even microscopic, decentralized energy supply options, leading the general public to conclude that it cannot compete as a substitute for the major conventional energy carriers. It is often represented as a kind of sticking plaster for the energy supply, which will have to be adopted to relieve the threatened environment but is really a second-class energy supply system, and an inevitable quasi last resort. This timidity has helped solar energy to appear to be only a marginal possibility and thus frequently at the bottom instead of the top of any list of measures dealing with energy policies, including the many suggestions for an environmental energy policy. Some committed nature conservationists believe that the defence of a microcosm is as important as the salvation of the macrocosm – and that the two aims are not necessarily compatible in every case.

For years, advocates of solar energy were told time and again that they should be modest in their demands because otherwise they would hurt 'their cause' – as if this were a favour to those demanding solar energy! They were warned to proceed very carefully, step by step, only after the most precise examination and always with a unanimous voice, as if solar energy were an unusually dangerous matter to be treated with the utmost caution. In stark contrast was the readiness and costly support given – and still being given – to really dangerous projects that produced the most expensive 'white elephants' imaginable.

Solar energy is only dangerous to the energy industry's interest in unchallenged continuity. How seriously it takes the alternative is indicated by the degree of opposition. For society, and for economic development, the opposite is true: the danger lies in restraining solar energy.

Champions of solar energy must be aware: motivation will not be strengthened by practising self-restraint and taking small steps, but by aggressive demand for big steps, and the claim to represent the all-encompassing alternative and not merely an additional element of energy supply. Such a claim would mark the transition from being one aspect of the old energy system to the concept of being a new energy system, bringing with it a new system of economy and society – the transition from 'solar tactics' to a solar strategy. Great alternatives are easier to mobilize than those that are too small. It is a widespread misconception that small political steps are generally easier to achieve than large ones. If the solar alternative is to succeed, the prevailing, tenacious small-mindedness surrounding vital issues must be overcome with the tools of social psychology. This, in turn, will generate demands and challenges to the political institutions to realign their priorities, not only for existing energy policies but, in general, along the lines of introducing solar energy – and no longer to see themselves as mere patrons, protectors and sponsors of the energy industry.

Those political institutions that have predominantly adopted the patterns of thought and action of the present energy industry are unable to fulfil their responsibilities to the energy system as a whole. Similarly, the political possibilities and responsibilities will be wasted if they are downgraded to a department-level responsibility. As long as decisions are being made on financial, agricultural, construction, transportation, technology, economic, local, development and foreign policies without taking the energy system into account, political institutions will be incapable of solving the essential questions. All individual sectors, departments and policies must consciously see themselves as links in the all-encompassing chain of alternative energy. Cabinets really should be energy cabinets in order to make certain that future decisions by all departments and at all levels are made for the benefit of a solar energy system that assures survival – and not just as an afterthought, but as a matter of top priority. A construction or agricultural ministry, for example, can make a greater immediate contribution to a new energy system than can an environment or energy ministry. A big city or region that accomplished the transformation to solar energy to the extent recommended here could start an avalanche by such an example and, with it, could cause political changes far beyond its own borders.

New strategies cannot be carried out by old methods. Part of a solar strategy must be a new understanding of political action, and new decision-making structures and processes that explode currently accepted frames of reference. Only in this way will political institutions be able to tap solar energy's potential. So far neither the political institutions, including parliaments, nor the political parties have achieved the required strategic and organizational level. Experience teaches that the upheaval needed to achieve practical, individual and institutional political action never comes about as the result of dialogue and rational insight (with some exceptions, but their room for manoeuvre remains limited as long as the other structures remain stuck in anachronisms). Without political propulsion from outside, far too little happens on the starting blocks of political responsibility. The political institutions that remain constrained by powerful special-interest groups must be brought round to the idea of repressing special interests that damage the common good, and actually assert the constitutional legitimacy that entitles them – and them alone – to make such decisions. Attempts to do this will succeed only if public support is mobilized. Such support can be enlisted by a public debate about energy that opens ears and reveals the contradictions, and not via the constantly invoked 'energy consensus' that expects society to deal reasonably with the representatives of the unreasonable, destructive energies.

The Movement for the New Century

Only increased public pressure and independent initiatives within society will ensure that solar energy unfolds fully in time. A new movement for a new century must arise to exert that kind of pressure with all the means provided by democratic constitutions – including the option of voting out of office those who refuse to turn the key to the door leading away from nature's destruction and towards new economic and social development.

The idealistic, spiritual and intellectual potential for such a movement exists now – much of it in the classical ideas contained in the major ideological currents, as they have crystallized in recent historical periods and have led to the creation of political parties. The problem is that these fundamental ideas are hardly identical any more with political reality as manifested by today's political parties, which have turned into conglomerates of more or less contradictory ideas and interests. The conventional wisdom that it is essential to demonstrate unity leads to the situation that the persistent forces of inertia continuously grind down any new initiatives to the point of unrecognizability. The official idealistic creeds are used to tie the hands of the truly concerned by those who no longer believe in alternatives, and who have made their peace with the dominating power constellations.

Anyone who prefers integration into the present energy system cannot point to any sort of hopeful future for the weakest – the growing number of unemployed, people in the developing countries and children, all of whom are facing devastated environments. At least those political parties that are based on notions of social improvement will not come through unchanged: a solar strategy is the opportunity to revive the ideal of social justice and the principle of self-determination. The social movement that emanated from the Industrial Revolution must no longer, as the Italian founder of the 'Manifesto' Group, Luciana Castellina, put it,

> limit itself to the search for new ways of more efficient production and more equitable distribution, of a larger amount of the same things, organized around the same product and consumption goals, but it must look for new technological possibilities to produce other things and, most of all, to arrive at another way of life.[190]

This requires a new material basis, and that is provided by a solar energy system. Solar energy is the pivotal point, for the transition from an 'irrational materialism'[191] to a rational materialism.

Experience teaches that the impulses and challenges necessary for this transition must come from a social movement, which is borne along by a practical idea and unleashes new strength and creativity. It is clear today – clearer than it was in the 1970s – that there exists an alternative capable of political and economic mobilization, and it is this basis that would provide strength for a socio-cultural movement on behalf of solar energy. It can set out a concrete and comprehensive alternative. Such a movement will be able to achieve the radical cultural change to a new spiritual and intellectual state of mind in society, from which new political initiatives would emerge, as well as new entrepreneurial points of departure, a new architecture and innumerable individual decisions for the use of energy, and all this without constant demands for financial help first.

In Switzerland, the Solar 91 effort has triggered a widespread citizens' movement, which wished to mark a historic turning point to a new age on the occasion of that republic's 700th anniversary.[192] This activism spurred many solar activities, especially at a local level. In Austria, a new grassroots movement is in the process of formation among ordinary citizens, who are constructing thousands of solar facilities with the help of local 'self-help groups'. In Germany, solar initiatives are springing up and organizing solar projects. In Denmark, there is widespread social support for renewable energies, stretching from the universities to farmers and manual workers. In particular the ironworkers' guild, which represents the installation trade, is involved. It is plain to see everywhere that once the alternative becomes part of public discussion and the outlook is widely known, sympathy and activities for the cause spread like wildfire. Popular referenda in some countries, such as Italy, have halted nuclear power use, at least domestically. If the political institutions do not adapt on their own to an ecologically based survival strategy, despite the practical possibilities, efforts should be made to initiate and conduct popular referenda to force the shift to solar energy by democratic means – for example, by imposing a 'solar penny' levy to finance society's conversion to solar energy.

In the long run, nobody will be able to prevent solar energy from supplanting other energy sources. As this book has shown, the others have nothing to offer in the long term. The current energy system, which is based on outdated forms of energy, is the most influential economic power on the globe, of course, but in fact it is already 'down' – the question is still open as to whether humanity will go 'down' with it, because it may succeed in maintaining the status quo – that is, the continuation of the nuclear/fossil energy system – for too long. The fateful question is whether solar energy will replace conventional energy in time! The longer the wait, the more certain it will be that the future will bring a cruel life of suffering for the vast majority of the world's growing population. And, if it is already doubtful now whether it is possible to summon up sufficient moral and political strength for the change, it is almost certain that it will not be later, as the menace of more destruction of the natural world and vanishing prospects of normal development for billions of human beings grows stronger. In other words, the more delay, the greater the effort needed for the salvation of civilization.

The thesis of the central importance of solar energy may be opposed by – justified – scepticism towards monocausal doctrines of salvation: it cannot be, it could be argued, that such comprehensive advantages for society could be achieved 'merely' with the use of solar energy. One could make the accusation that this is nothing but another false attempt to solve social problems with a new technology. But it is precisely the basic importance of energy and the laws of thermodynamics that assign this significance to renewable energies. It is not the arguments in this book that are 'monocausal', but the sun, the central star, that is monocausal for the ecosphere. And it is not just a particular technology that provides this broad perspective but precisely this solar power, whose overwhelming importance for this planet is an undisputable matter of fact. The challenge is to attune the planet's modern technology to the laws of the sun. And because over the ages the technical capabilities have developed to meet that challenge, even for a growing human race, mankind has a unique opportunity – and precisely at the time of its most serious collective danger.

The utilization of solar energy is in no way a simple solution to complex problems. The crude answer consisted, and still consists, in the use of fossil and nuclear energies and in increasingly uniform processes in agriculture and industry, which have led to a destruction of natural diversity inappropriate to nature's complexity. Rather, solar energy use means to adapt human civilization once more to the intrinsic forms of nature.

A peace treaty between mankind and nature is not possible without a global solar energy economy. We are very pressed for time but, on the other hand, we have the opportunities for a solar strategy within our grasp and have reached the point where we should no longer bear with the inhuman delaying tactics of those who 'don't do what they know must be done'. To use the 'energy of the people', we need to mobilize the people's energies – to stir up a solar energy revolution.

There is no alternative.

Notes

1   Fritjof Capra, *The Turning Point: Science, Society, and the Rising Culture.* New York: Simon & Schuster, 1982.

2   Hans Jonas, *Der Spiegel,* No. 20/1992, pp. 92–95.

3   Carl Amery, *Natur als Politik: Die ökologische Chance des Menschen* (Nature as Politics: The Ecological Chance for Man). Reinbek/Hamburg: Rowohlt, 1976, pp. 92–94.

4   Robert Kurz, *Der Kollaps der Modernisierung: Vom Zusammen-bruch des Kasernensozialismus zur Krise der Weltökonomie* (The Collapse of Modernization: From the Breakdown of Barracks' Socialism to the Crisis of the World Economy). Frankfurt/Main: Eichborn, 1991, p. 270.

5   Wilhelm Ostwald, *Energetische Grundlagen der Kultur-wissenschaft* (Energetic Foundations of the Cultural Sciences). Leipzig: W. Klinkhardt, 1909, pp. 41–44.

6   Christian Schütze. *Das Grundgesetz des Niedergangs* (The Basic Law of Decline). Munich, 1989.

7   Enrico Turrini, *La Via del Sole* (The Way of the Sun). Fiesole: S. Domenico di Firenze, 1990, p. 11.

8   Christian Meier, *The Greek Discovery of Politics.* Cambridge, Mass.: Harvard University Press, 1990.

9   Dolf Sternberger, *Die Politik und der Friede* (Politics and Peace). Frankfurt/Main: Suhrkamp, 1986, p. 76.

10 Oskar Negt and Alexander Kluge, *Maßverhältnisse des Politischen* (Measurement Proportions of Politics). Frankfurt/Main: Fischer, 1992, p. 17.

11 Serge Podolynski, 'Menschliche Arbeit und Einheit der Kraft' (Human labour and unity of power), *Die Neue Zeit,* March/April 1883, pp. 449–451.

12 Juan Martinez-Alier, *Ecological Economics: Energy, Environment and Society.* Oxford, 1987.

13 Jean-Claude Debeir, Jean-Paul Deléage and Daniel Hémery, *In the Servitude of Power: Energy and Civilization through the Ages.* London-New Jersey: Zed Books, 1991 (Paris, 1986), pp. 3–24, 100–125.

14 Hartmut Elsenhans, *Migration und Wirtschaftsentwicklung* (Migration and Economic Development). Frankfurt/Main-New York: Campus, 1978, pp. 15–30.

15 *Atlas Mondial de L'Energie* (World Energy Atlas). Paris, 1989, pp. 44–45.

16 The data in Table 2 have been taken from the *Human Development Report 1992* of the United Nations Development Programme (UNDP), pp. 130–173.

17 Earl Cook, 'The flow of energy in an industrial society', *Scientific American,* no. 9/1971, p. 136.

18   Paul A. Ryan, 'Some policy and economic realities of biomass development and management in Africa and Asia', in Michael A. Kuliasha, Alexander Zucker and Kerry J. Bellew, *Technologies for a Greenhouse-Constrained Society*. Oak Ridge, 1991, p. 548.

19   John G. Clark, *The Political Economy of World Energy*. University of North Carolina Press, 1990, p. 4.

20   Cited in: Daniel Yergin, *The Prize: The Epic Quest for Oil, Money and Power*. New York: Simon & Schuster, 1991, p. 401.

21   Hartmut Elsenhans, *Erdöl für Europa* (Oil for Europe). Hamburg, 1979, pp. 11–13.

22   Mohssen Massarrat, *Weltenergieproduktion und Neuordnung der Weltwirtschaft*. Frankfurt, 1990, p. 169.

23   Clark (see note 19), p. 134.

24   Mohssen Massarrat, *Überfluss trotz Erschöpfbarkeit* (Abundance in spite of Exhaustibility). Manuscript. Osnabrück, 1991, pp. 38-39.

25   Yergin (see note 20), p. 770.

26   Hefez Sabat, *Die Schuld des Nordens* (The Debt of the North). Bad Konig, 1991, pp. 38ff.

27   Mark Kosmo, *Money to Burn? The High Costs of Energy Subsidies*. Washington DC: World Resources Institute, 1987, pp. 8–9 and 34.

28   Ian Smart, 'Energy and the power of nations', in Daniel Yergin and Martin Hillenbrandt, *Global Insecurity*. New York: Penguin, 1982, pp. 349–374.

29   J. Goldemberg, Th.B. Johannson, A.K.N. Reddy and R.H. Williams, *Energy for a Sustainable World*. Washington DC: World Resources Institute, 1987, pp. 103–104.

30   Quoted in: Charles A. Bane, *The Electrical Equipment Conspiracies (The Treble Damage Actions)*. New York: Federal Legal Publications Inc., 1973, vol. 1, p. 14.

31   Hans Kronberger, Blut für Öl (Blood for Oil). Wien: Uranus, 1998.

32   Uwe Holtz 'Den Ökozid verhindern!' (Prevent the ecocide!), in Hermann Scheer (ed.), *Das Solarzeitalter* (The Solar Age). Karlsruhe, 1989, pp. 94–107. C.F. Müller

33   The Harare Solar Energy Declaration, 17 November 1991, in *The Yearbook of Renewable Energies*. Bochum: Ponte Press, 1992, pp. 234–235.

34   Klaus Bosselmann, *Im Namen der Natur* (In the name of nature), München, 1992, p. 24–25).

35   Paul Kennedy, *The rise and fall of great powers*, NY, 1987.

36   Wilhelm Ostwald, *Der Energetische Imperativ* (The Energy Imperative). Leipzig, 1912, pp. 83–86. Verlag: Dr.Klinkhardt.

37   Nicolas Georgescu-Roegen, *The Entropy Law and the Economic Process*. Cambridge, Mass.: Harvard University Press, 1971.

38   Barry Commoner, *The Poverty of Power: Energy and the Economic Crisis*. New York: Random House, 1976, p. 213.

39   Schütze (see note 6), p. 19.

40   Peter Kafka, *Das Grundgesetz vom Aufstieg* (The Basic Law of Ascendancy). Munich-Vienna: Hanser, 1989, p. 81.

41   Georges Alexandroff and Alain Liébard, *L'Habitat Solaire* (Solar Radiation). Comment (Commentary). Paris, 1979, p. 10.

42   Helmut Tributsch, *Rückkehr zur Sonne* (Returning to the Sun). Berlin, 1979, p. 109.

43   Augustin Mouchot, *Die Sonnenwärme und ihre industriellen Anwendungen* (original title: *La Chaleur Solaire et ses Applications Industrielles*). In English: *The Heat of the Sun and its Industrial Applications*. Oberbözberg, Switzerland: Olynthus Verlag, 1987 (new edition).

44   Adolf Goetzberger, 'Die Anwendungen verbessern: Forschung und Entwicklung auf dem Weg ins Solarzeitalter' (Improving the applications: research and development on the way to the solar age), in Scheer (see note 32), p. 57.

45   Farrington Daniels, *Direct Use of the Sun's Energy*. New York, 1974, pp. 6–8.

46   Erich Hau, *Windkraftanlagen* (Wind Power Plants). Heidelberg, 1989, pp. 22–34. Springer.

47   Wolfgang Palz, *Biogasanlagen in Europa* (Biogas Plants in Europe). Köln, 1985, pp. 1–2.

48   Marcel Perrot, *La houille d'or* (The Golden Coal). Paris: Fayard, 1963.

49   Jurgen Kleinwächter, 'Die COMPLES', *Sonnenenergie*, vol. 3, no. 3/1978, pp. 20-21.

50   Werner v Siemens, 'Über die von Hm. Fritts in New York entdeckte elektromotorische Wirkung beleuchteter Selens' (On the electric-motor effect of illuminated seleniums), *Monatsberichte der Berliner Akademie der Wissenschaften vom 3 Mai 1875 und 7 Juni 1877* (Monthly Reports of the Academy of Science in Berlin dated 3 May 1875 and 7 June 1877).

51   Ostwald (see note 5), p. 44.

52   *The Sun in the Service of Mankind*. Paris: CNRS, 1973.

53   Wolfgang Palz, *Solar Electricity: An Economic Approach to Solar Energy*. London, 1978, pp. 116ff.

54   Barry Commoner, *The Politics of Energy*. New York: Knopf, 1979, p. 36.

55   Quoted in: Modesto A. Maidique, 'Solar America', in *Energy Future: Report of the Energy Project at the Harvard Business School* (eds Robert Stobaugh and Daniel Yergin). New York, 1979, p. 193. Random House.

56   Council on Environmental Quality, Executive Office of the President: *Solar Energy, Progress and Promise*, April 1978.

57   Maidique (see note 55), pp. 188 and 212.

58   Le Groupe de Bellevue. Paris, 1978.

59   *Report of the United Nations Conference on New and Renewable Sources of Energy*. New York: United Nations, 1981 (Doc. A/Conf 100/11).

60   *R&D Expenditures in the Field of Space Technology. Analysis by the Scientific Service of the German Bundestag*, 10 March 1992.

61 Hermann Scheer, *Die Befreiung von der Bombe* (Liberation from the Bomb). Köln: Bund, 1986, pp. 292–298.

62 Barbara Tuchman, *March of Folly: From Troy to Vietnam*. New York: Knopf, 1984, pp. 4 and 387.

63 Christine and Ernst-Ulrich von Weizsacker, 'Fehler-freundlichkeit und Evolutionsprinzip' (High error rates and the principle of evolution), *Universitas*, no. 41/483, 1986, pp. 791–798.

64 Münchner Rückversicherung (Munich Reinsurance Inc.), Sturm. (Storm) Munich, 1990.

65 Ben-Alexander Behnke, *Abschied von der Natur*. Düsseldorf, 1997.

66 Jesco von Puttkamer in the film *Gestern war heute noch morgen*, Wendländische Filmkooperative, 1991.

67 Kafka (see note 40).

68 Inquiry Commission of the 11th German Bundestag (ed.), *Preventive Measures to Protect the Earth's Atmosphere*. Third Report, vol. 1, Bonn, 1990.

69 World Resources Institute (see note 29), p. 19.

70 Inquiry Commission 'Protecting the Earth's Atmosphere' of the German Bundestag (ed.), *Climate Change: A Threat to Global Development*. Bonn, 1992, pp. 173–174.

71 Robert Repetto, *Promoting Environmentally Sound Economic Progress: What The North Can Do*. Washington DC: World Resources Institute, April 1990, pp. 13–14.

72 *FORATOM-Newsletter*, no. 19, Brussels, June 1992, pp. 5–6.

73 Eckehard Rebhan, *Heißer als das Sonnenfeuer. Plasmaphysik und Kernfusion* (Hotter than the Fire from the Sun. Plasma Physics and Nuclear Fusion). München: Piper, 1992, p. 445.

74 *R&D expenditures for nuclear fission and fusion*. Analysis by the Scientific Service of the German Bundestag, 4 March 1992.

75 Tributsch (see note 42), p. 54.

76 D. Pfirsch and K.H. Schmitter, *On the Economic Prospects of Nuclear Fusion with Magnetically Confined Plasmas*. Max-Planck-Institut für Plasmaphysik, Report-Nr. IPP 6/271, December 1987.

77 Jochen Benecke, 'Kernfusion ist keine Alternative' (Nuclear fusion is no alternative), *Bild der Wissenschaft*, no. 2/1987, p. 128.

78 Rebhan (see note 73), p. 444.

79 Ibid, p. 448.

80 Cesare Marchetti, 'Nach der Kernkraft kommt die Kernfusion' (Nuclear fission is followed by nuclear fusion), *Bild der Wissenschaft*, no. 8/1988, pp. 111–113.

81 Harry Lehmann, 'Nuclear fusion: energy policy option or scientific playground', in *Yearbook of Renewable Energies 1992*. Bochum: Ponte Press, 1992, p. 95.

82 Benecke (see note 77).

83 Quoted in: Jochen Benecke, 'On the prospect of power reactors based on nuclear fusion', paper prepared for the Scientific and Technological Options Assessment (STOA) of the European Parliament, February 1988.

84 Jeremy Rifkin, *Entropy: A New World View*. London: Paladin, 1985, pp. 123 and 125.

85 Josef Spitzer, 'Weltweite Verfügbarkeit der Energieträger von Heute und Morgen' (Global availability of today's and tomorrow's energy carriers), in W. Pillmann and S. Burgstaller (eds), *Proceedings of the International Forum of the Environment*, 23–26 September 1990, Bad Kleinkirchheim, Austria. Vienna: International Society for Environmental Protection, 1991.

86 EUROSOLAR / European Solar Council, Entwicklung und Arbeitsplatzpotential Erneuerbarer Energien in der Europäischen Union. *Solarzeitalter* No. 3 / 1997, pp. 11–31.

87 *The Potential of Renewable Energy. An Interlaboratory White Paper.* Colorado: Solar Energy Research Institute, March 1990.

88 Irm Pontenagel (ed.), *Das Potential erneuerbarer Energien in der Europäischen Union. Ansätze zur Mobilisierung erneuerbarer Energien bis zum Jahr 2020*. Heidelberg: Springer, 1995.

89 Hermann Scheer, '100% Renewable Energies', *The Yearbook of Renewable Energies 1995/96*. London: James & James (Science) Publishers, 1995, p. 1.

90 Hans Schnitzer, 'Technical and Economic Fundamentals for an Exclusively Solar-Based Energy System', *The Yearbook of Renewable Energies 1995/96*. London: James & James (Science) Publishers, 1995, pp. 86–93.

91 Phillip Elliot and Roger Booth, 'Sustainable biomass energy', Shell: Selected Papers, 1991.

92 Karl-Peter Hasenkamp, 'Global reforestation to solve the problem of CO_2 pollution', in *The Yearbook of Renewable Energies 1992*. Bochum: Ponte Press, 1992, pp. 96–101

93 United States Environmental Protection Agency, 'Response and Feedback of Forest Systems to Global Climate Change', September 1990.

94 R.A. Houghton and G.M. Woodell, 'Global climatic change', *Scientific American*, no. 4/1989, p. 26.

95 Frank Rosillo Calle and David O. Hall, 'Biomass energy, forests and global warming', *Energy Policy*, vol. 20, no. 2/1992, pp. 124–136.

96 *Towards Sustainable Energy Development: The Energy Activities of the UN System and the Development Banks*. Stockholm: The Beijer Institute, 1989.

97 Richard St Barbe Baker, *Man of the Trees*. New Delhi: Indian National Trust for Art and Cultural Heritage, 1989.

98 Dennis Anderson and Catherine D. Bird, 'Carbon accumulations and technical progress', manuscript, University College London and Balliol College, Oxford, December 1990, p. 19.

99 Ibid, p. 22.

100 EUROSOLAR, Memorandum: Amerikanische und japanische Aktivitäten für erneuerbare Energien, Bonn, 1997.

101 Ibid

102   For more information see: 'Solar energy: a strategy in support of environment and development. A comprehensive analytical study on renewable sources of energy', Report of the UNSEGED (United Nations Solar Energy Group on Environment and Development), in *The Yearbook of Renewable Energies 1992*. Bochum: Ponte Press, 1992, pp. 5–34; R. Howes and A. Feinberg (eds), *The Energy Sourcebook*. New York: American Institute of Physics, 1991; Thomas B. Johannson, Henry Kelly, Amulya K.N. Reddy and Robert H. Williams (eds), *Renewable Energy: Sources for Fuels and Electricity*. Washington DC: Island Press, 1993, pp. 1–15; Commission of the European Communities, Programme information no. 523, 24 July 1992; G.T. Wrixon, A.-M.E. Rooney and W. Palz, *Renewable Energy: 2000*. Berlin-Heidelberg: Springer, 1993.

103   W.B. Gillet and J.R. Stammers, *Review of Active Solar Technologies*. Final Report for the Energy Technology Support Unit, UK, 1992.

104   G. Long, *Solar Aided District Heating Systems in the UK*. Report for the Energy Technology Support Unit, UK, 1992.

105   E. Gruber, H. Erhorn and J. Reichert (eds), *Solarhauser Landstuhl* (Landstuhl Solar Houses). Köln (Cologne), 1989.

106   Sibylle and Jörg Schleich, *Erneuerbare Energien nutzen* (Using Renewable Energies). Düsseldorf, 1991, pp. 155–112.

107   J. Benemann and R. Aringhoff, 'The technological potential of solar thermal power generation', in *The Yearbook of Renewable Energies 1992*. Bochum: Ponte Press, 1992, pp. 44–52.

108   *Systemvergleich und Potential von solarthermischen Anlagen im Mittelmeerraum* (Comparison of Systems and the Potential of Solar Thermal Power Installations in the Mediterranean Basin). Study by DLR (Deutsche Forschungsanstallt für Luft- und Raumfahrt), ZSW (Zentrum für Sonnenenergie- und Wasserstoffor-schung) Stuttgart, Interatom and SBP (Schlaich, Bergermann and Partner), Stuttgart, 1992.

109   EUREC-Agency: PV for the World's Villages, Feasibility Study for the EU-Commission. November 1996.

110   R. Hill *et al.*, *The Potential Generating Capacity of PV-Clad Buildings in the UK*. Newcastle Photovoltaics Applications Centre, May 1992.

111   *Proceedings of the Eighth EC Photovoltaic Solar Energy Conference*, Florence, Italy, 9–13 May 1988, pp. 149–155.

112   Harry Muuß, 'Der Beitrag der Windenergieversorgung zu einer regenerativen Energiewirtschaft' (The contribution of wind energy in a renewable energy economy), in H. Scheer (ed.), *Die gespeicherte Sonne* (The Stored Sun). Munich: Piper, 1987, pp. 229–256.

113   F. Rosillo-Calle and D.O. Hall, 'Brazilian alcohol: food versus fuel?', *Biomass*, vol. 12, pp. 97–128.

114   For cropland, see: D.O. Hall, F. Rosillo-Calle, R. Williams and J. Woods, 'Biomass for energy: supply prospects', in Johansson *et al.* (see note 102), pp. 640–641. For degraded land, see: A. Grainger, *International Tree Crops Journal*, No. 5/1988, p. 31.

115   Cynthia Pollock Shea, *Renewable Energy: Today's Contribution, Tomorrow's Promise*, World Watch Paper 81. Washington DC: World Watch Institute, January, 1988, pp. 18–26.

116 David Wright, *Biomass: A New Future?* Commission of the European Communities, Forward Studies Unit, 1991, p. 34.

117 Giuliano Grassi, 'Biomass research and development activities in the EC', in *The Yearbook of Renewable Energies 1992*, Bochum: Ponte Press, 1992, pp. 63–64.

118 Stig Ledin, *Willow as a Biomass Fuel Resource*. Uppsala: Swedish University of Agricultural Sciences, Section for Short Rotation Forestry, 1972.

119 Wolfgang Ständer, *Neue ökonomische Lösungen für die Agrarüberproduktion; die Energie-, die Umweltprobleme und die Erzeugung von kompostierbaren Industrie-produkten aus C4-Schilfpflanzen* (New Economic Solutions for Agricultural Overproduction: Energy and Environmental Problems and the Production of Compostable Industrial Products made of C4-Reeds). Munich: Polytechnisches Institut, 1992; and *Franz Alt, Schilfgras statt Atom* (Reeds instead of Atoms). Munich, 1992. Piper

120 Shea (see note 115), p. 23.

121 Roland Emrich, 'Reishülsenals Brennstoff' (Rice husks as fuels), *Solarzeitalter*, No. 2/1991, pp. 6–10.

122 L.-A. Kristofersen and V. Bokalders, *Renewable Energy Technologies: Their Applications in Developing Countries*. Southampton, 1991, p. 29.

123 H. Schäffler, *Das Potential der Biomasse* (The Potential of Biomass). Stuttgart, 1992. See also: G. Grassi, G. Trebbi and D.C. Pike (eds), *Electricity from Biomass*. Brussels, 1992.

124 Asbjorn Vinjar, 'Hydropower: utilization of waterfall energy in an environmentally sound, sensitive world', in *The Yearbook of Renewable Energies 1992*. Bochum: Ponte Press, 1992, pp. 82–86.

125 Henry Kalb and Werner Vogel, 'Solarstrom und Solarwasserstoff' (Solar electricity and solar hydrogen), in Scheer (ed.) (see note 112), pp. 213–227.

126 Othmar Heise, 'Zur Methodik der Schadens-minimierung: em Wegbereiter für erneuerbare Energien' (Methodology of minimizing damage: a path for renewable energies), *Solarzeitalter*, No. 4/1991, pp. 16–26.

127 Werner Freiesleben. Beginnen wir mit der Eisenbahn (Let's Start Like the Railways), in Scheer (ed.) (see note 32), pp. 38–68.

128 Joachim Nitsch, *Forschungs- und Entwicklungsbedarf der solaren Energietechnik* (Research and Development Needs for a Solar Energy Technology). Stuttgart: DLR-Programmstudie, 1992.

129 Jurgen Kleinwächter, *Vorschlag eines Deutschen Thermosolaren Forschungszentrums* (DTF) (Proposal for a German Solar Thermal Research Centre). Study by Bomin-Solar, Lorrach, 1990.

130 Helmut Tributsch, 'Die Solarzelle: Schlüsselelement für die Wirtschaftlichkeit solaren Wasserstoffs' (The solar cell: a key element for the economic competitiveness of solar hydrogen), in Scheer (ed.) (see note 112), p. 121.

131 'EUROSOLAR-Memorandum zur Einbeziehung der Solarenergie in die Grundlagenforschung' (EUROSOLAR Memorandum for Integrating Solar Energy into Basic Research), *Solarzeitalter*, No. 2/1992, pp. 15–16.

132 Max Weber, 'Energetische Kulturtheorien' (Energetic cultural theories), in *Gesammelte Aufsatze zur Wirtschaftslehre* (Collected Papers on Economic Studies). Tubingen, 1968, p. 409. Mohr/Siebeck.

133 Elmar Altvater, *Vom Wohlstand und Mißstand der Nationen* (Wealth and Disgrace of Nations). Münster, 1982. Westfölisches Dampfboot.

134 Hazel Henderson, *Creating Alternative Futures: The End of Economics*. New York: Berkley, 1978, p. 83.

135 Egon Matzner, *Wohlfahrtsstaat und Wirtschaftskrise* (Welfare State and Economic Crisis). Reinbek/Hamburg: Rowohlt, 1978, p. 81.

136 Fredmund Malik and Daniel Stelter, *Krisengefahren in der Weltwirtschaft: Überlebensstrategien fur das Unternehmen* (Danger of Crises in the World Economy: Survival Strategies for Companies). Stuttgart, 1990, pp. 20–22.

137 Olav Hohmeyer, 'The social cost of electricity: renewables versus fossil and nuclear energy', *International Journal of Solar Energy*, vol. 11, 1992.

138 Nicolas Georgescu-Roegen, 'The entropy law and the economic process in retrospect', *Eastern Economic Journal*, vol. XII, no. 1, Jan–Mar 1986, p. 16.

139 Wolfgang Pala and Henri Zibetta, 'Energy pay-back time of photovoltaic modules', *International Journal of Solar Energy*, vol. 10, 1991, pp. 211–216.

140 J. Schmid and H.P. Klein, *Performance of European Wind Turbines*. London-New York, 1991, p. 134.

141 Elmar Altvater, *Die Zukunft des Marktes* (The Future of the Market). Münster, 1991, p. 93. Westfölisches Dampfboot.

142 e.g. Wright (see note 116), p. 10.

143 William W. Kaufmann and John D. Steinbrunner, *Decisions for Defense*. Washington, 1991, p. 8.

144 Following on information of the US Department of Energy.

145 Harold M. Hubbard, 'The real cost of energy', *Scientific American*, April 1991, pp. 19–20.

146 Commoner (see note 54), p. 44.

147 Commoner (see note 38), pp. 131–132.

148 Gottfried Rössle, *Das Maren-Modell: Perspektiven einer Energiezukunft* (The Maren Model. Perspectives for the Future of Energy). Hof, 1989, pp. 175–177.

149 Max Planck, *Wissenschaftliche Autobiographie* (Scientific Autobiography). Leipzig, 1928, p. 22.

150 Thomas S. Kuhn, *The Structure of Scientific Revolutions*. Chicago, 1962, pp. 168 and 158. University of Chicago Press.

151 John Keyes, *The Solar Conspiracy*. New York, 1975.

152 Daniel M. Berman, John T. O'Connor, *Who owns the sun?* White River Junction, 1996.

153 Lutz Mez, RWE. *Geschichte eines Konzerns*, Köln, 1996.

154 Hendrik Paulitz, *Manager der Klimakatastrophe*, Göttingen, 1994.

155 H.K. Schneider and W. Schulz, *Investment Requirements of the World Energy Industries 1980–2000*. Cologne, 1980.

156 Massarrat (see note 24), p. 63.

157 Christian Meier, *Die Ohnmacht des allmächtigen Diktators Caesar* (The Powerlessness of the All-Powerful Dictator Caesar). Frankfurt/Main, 1980, pp. 13–14.

158 Hermann Scheer, *Zürück zur Politik*, Bochum, PontePress, 2. Auflage, 1998.

159 Martin Jaenicke, *Staatsversagen: Die Ohnmacht der Politik in der Industriegesellschaft* (Failure of the State: The Impotence of Politics in the Industrial Society). Munich, 1986. Piper.

160 Klaus Heinloth, *Die Energiefrage*. Wiesbaden, Vieweg, 1997.

161 Tributsch (see note 42), pp. 188-190.

162 Ernst Ulrich von Weizsäcker, *Erdpolitik*. Wiesbaden, 1989, p. 78.

163 Hans-Christoph Binswanger 'Die verlorene Unschuld der Windenergie', *Blätter für deutsche und internationale Politik*. Nr. 10 / 1997. pp. 1272–1275.

164 Holger Strohm, *Friedlich in die Katastrophe* (Peacefully towards the Catastrophe). Frankfurt/Main, 1981, pp. 1–43.

165 Manfred Eigen, *Jenseits von Ideologien und Wunschdenken: Perspektiven der Wissenschaft* (Beyond Ideologies and Wishful Thinking: Perspectives of Science). Munich-Zürich, pp. 243–249.

166 Irm Pontenagel, 'Was sind 'additive Energien?': Eine grundsätzliche Betrachtung über Semantik und Paradigmen in der Energiepolitik' (What are 'additive energies?': general reflections on semantics and paradigms in energy policy), *Solarzeitalter*, No. 3/1991, p. 1.

167 EUROSOLAR, *Solar Powered Boats for Venice*. Bonn, 1995.

168 Peter Hennicke (Ed.), *Den Wettbewerb im Energiesektorplanen: Least-Cost-Planning als neue Methode zur Optimierung von Energiedienstleistungen* (Planning Competition in the Energy Industry: Least-Cost-Planning as a New Method for Optimizing Energy Services). Heidelberg, 1991.

169 EUROSOLAR, *EURERULE*. Bonn, 1996.

170 See: Willy Leonhardt, Reinhard Klopffleisch and Gerhard Jochum, *Kommunales Energiehandbuch* (Energy Handbook for Communities). Karlsruhe, 1989; Siegfried Rettich, *'Das Rottweiler Modell'* (The Rottweil Model), *Solarzeitalter*, No. 2/1992, D. l.

171 Hermann Scheer, Erneuerung der Kohleregionen durch erneuerbare Energien, EUROSOLAR-Paper, 1997.

172 Kurt Müller, 'The eco-pyramid: criteria for financing solar projects', in *The Yearbook of Renewable Energies 1992*. Bochum: Ponte Press, 1992, pp. 198–202.

173 Michael T. Eckart, Maurice E. Miller, Die Solar-Bank, *Solarzeitalter*, No, 3 / 1997, pp. 6–10.

174 William Pascoe Watkins, *The International Co-operative Alliance*. England, 1970.

175 *Economic and Fiscal Instruments of Environment Policy*. Report of the Committee on the Environment, Public Health and Consumer Protection of the European Parliament (Rapporteur: Manfred Vohrer), EP-Document A3-0130/91, dated 13 March 1991. The Report contains an overview on most of the proposals being discussed in the scientific community. For another overview of the whole debate, see: Reinhold Buttgereit and Peter Palinkas, *Internationalization of External Effects in Energy and Environment Policy, their Incorporation into National Accounting Systems and the Debate Surrounding a CO_2 and Energy Tax in the European Community*. European Parliament Working Paper, Energy and Research Series W1, March 1992.

176 Ernst-Ulrich von Weizsäcker, 'Keine Angst vor hohen Energiepreisen' (Don't be afraid of high energy prices), *EUROSOLAR-Journal*, No. 2/1990, pp. 32–34.

177 Malcolm Slesser, *UNITAX: A New Environmentally Sensitive Concept in Taxation*. Edinburgh: Resource Use Institute, 1989.

178 Helge Hveem, 'Minerals as a factor in strategic policy and action', in Arthur H. Westing (ed.), *Global Resources and International Conflict*. Oxford, 1986, pp. 55–113.

179 Ugo Bilardo and Giuseppe Mureddu, 'Analisi e prospettive dell' attività del riciclo dei minerali' (Analysis and forecasts of minerals recycling activity), *Energie e Materia Prime*, No. 33-34/1983.

180 Reimund Bleischwitz and Helmut Schütz, *Unser trügerischer Wohlstand* (Our Deceptive Wealth). Wuppertal Institute for Climate, Environment and Energy, 1992, pp. 24–27.

181 J. Gretz, W. Korf and H. Lyons, 'Wasserstoff in der Stahlindustrie' (Hydrogen in the steel industry), *Solarzeitalter*, No. 4/1990, pp. 20–22.

182 J. Butlin, 'Enhancing recycling through a material tax', *Resources Policy*. Sep. 1983.

183 Ugo Bilardo and Giuseppe Mureddu, *Energy, Raw Materials for Industry and International Cooperation*. Rome: ENEL, 1989, p. 34.

184 'EUROSOLAR Memorandum for the Establishment of an International Solar Energy Agency (ISEA)', in *The Yearbook of Renewable Energies 1992*. Bochum: Ponte Press, 1992, pp. 213–219.

185 Franz Vranitzky, 'Österreichische Energiepolitik' (Austrian energy policy), *Solarzeitalter*, No. 2/ 1992, p. 7.

186 'Die Brennholzkrise' (The firewood crisis), UNESCO-Kurier, No. 1/1989.

187 Kurt Egger and Bernhard Glaeser, 'Ideologiekritik der Grünen Revolution' (Ideological criticism of the green revolution), in *Technologie und Politik 1*. Reinbek/Hamburg: Rowohlt, 1977, pp. 135–136.

188 Peter de Groot, Alison Field-Juma and David O. Hall, *Taking Root: Revegetation in Semi-Arid Kenya*. Nairobi-Harare, 1992.

189 Karl-Peter Hasenkamp and Hermann Scheer. 'Large-scale international program for afforestation of large countries as a contribution to warding off a climate change', in *The Yearbook of Renewable Energies 1992*. Bochum: Ponte Press, 1992, pp. 185–189.

190 Luciana Castellina, 'why 'red' must be 'green' too', in Miklos Nikolic (ed.), *Socialism on the Threshold of the Twenty-first Century*. Belgrade, 1985, p. 56.

191 Amery (see note 3), pp. 17–19.

192 Gallus Cadonau, 'Solar 91: for more Swiss self-sufficiency in energy', in *The Yearbook of Renewable Energies 1992*. Bochum: Ponte Press, 1992, pp. 105–114.